About the Author

Larry L. Hench is Professor of Ceramic Materials at Imperial College of Science, Technology and Medicine, and Co-Director of the Imperial College Tissue Engineering Centre. A world-renowned scientist, he graduated from the Ohio State University and is a Member of the National Academy of Engineering, USA. His eminent career spans 40 years and several fields, including biomedical materials, tissue engineering, optical materials and safe disposal of radioactive waste. His scholarly writings — 500 research papers and 25 books — and 27 patents have led to numerous awards and honours. He is credited with the discovery of Bioglass®, the first man-made material to bond to living bone — benefiting millions of people world-wide. Professor Hench's concern about education and moral issues extends to the children's books written by him, which feature Boing-Boing the Bionic Cat.

Science, Faith and Ethics

Larry L. Hench

Imperial College

Imperial College Press

Published by

Imperial College Press
57 Shelton Street
Covent Garden
London WC2H 9HE

Distributed by

World Scientific Publishing Co. Pte. Ltd.
P O Box 128, Farrer Road, Singapore 912805
USA office: Suite 1B, 1060 Main Street, River Edge, NJ 07661
UK office: 57 Shelton Street, Covent Garden, London WC2H 9HE

British Library Cataloguing-in-Publication Data
A catalogue record for this book is available from the British Library.

ISBN   1-86094-219-9
ISBN   1-86094-220-2 (pbk)

Printed in Singapore by World Scientific Printers (S) Pte Ltd

Contents

Foreword

To introduce this book, *Science, Faith and Ethics*, is a double privilege. The Chautauqua Institution is an educational and cultural centre in the state of New York in the USA. As director of the Institution's Department of Religion in 1991, I invited Professor Larry Hench to give five addresses in the Hall of Philosophy. He chose as his theme "Science, Faith and Ethics". Larry Hench is a scientist who is also a believer. He does not live in two worlds separated by a chasm centuries in the making. He is a single, unitary being, living and working in a world that, as scientist, he investigates and that, as believer, he acknowledges to be created by God, the God who is maker of all things in heaven and earth. To read these 1991 lectures, now further reflected on, worked out and refined, is to acknowledge with gratitude the workings of a mind that has sought throughout a notable career to be true to the demands and disciplines of both modes of thinking, the scientific and the theological.

My second occasion for gratitude is that the lectures, now in this richer and fuller form, are a lucid commentary on what one of the founders of the Chautauqua Institution over a century ago called "the Chautauqua idea". This is, he said, the idea that "all life is one and religion belongs everywhere".

The words come from a form of late 19th century Protestant piety, but they are entirely appropriate to the emerging 21st century conversation between scientists and theologians.

From the time of the Enlightenment, science and religious faith developed two increasingly independent ways of viewing, speaking about and describing the core of their disciplines. The two languages and their separate descriptions of reality seldom intersected. To illustrate: At a June 2000 conference held in Philadelphia and funded by the John Templeton Foundation, scientists and theologians assembled to address the intriguing theme "Extending Life, Eternal Life".

Evolutionary biologists present at the conference argued, and with some cogency, that human beings are within touching distance of being able to control their own evolutionary destiny and to extend the average life span far beyond its present day limits. Some even argued that in future we may come to regard death itself as a disease to be cured rather than an inevitable consequence of cell death.

In contrast, the theologians at the conference spoke of eternal life, but they did so largely in spiritual terms, defining eternal life as a way of being in relation to God and living towards God. The impression I had in listening to both was of watching and hearing two parallel tracks of speakers, with those on one track using the language of science and those on the other the language of theology, but neither entering significantly into genuine dialogue with the other.

The Philadelphia conference was a clear example of the split between the "How?" of science and the "Why?" of the believer. The dichotomy is typical of many conversations between scientists and religious believers. This dangerous split creates a division between God and the world and eventually leads to a denial of any interaction between God and the world and to the claim that nothing lies beyond the explanations given by the physical sciences. Accepting the split, the scientist will then pursue research into the truth of his or her world of study, but will do so without reference to the other voice in the quest for truth, that of the theologian and the religious believer.

The religious believer, in a similar way, will continue to speak about God or eternal life, but without reference to the world and to the voice of the scientist.

The time has come when it is important for scientists to think intentionally about the unthought in their way of looking at the core of their disciplines, and for theologians likewise to think intentionally about the unthought in their own way of looking at their doctrines, ethical teachings, rites, worship, prayers and ceremonies.

The question that theologians can ask, in conversation with scientists, is: "If the world had a beginning, who made that beginning, and what does it mean that we have a universe at all?"

There is an unthought in science that needs to be thought about. Correspondingly, the question that scientists can ask, in conversation with theologians, is: "How do you connect what you say about the divine Creator or about God's revelation, with what we have to say in our physics, chemistry, biology, sociology and psychology?"

So scientists can reflect and say, "Yes, when we use our instruments to measure some part of the empirical world, we must also recognize that we have no way of measuring fully what appears to be a unitary and intelligible cosmos of contingency and freedom."

There is an unthought in theology that needs to be thought about. Theologians can reflect and say, "Yes, when we speak about God as Creator we must acknowledge that God does not create as we do, and therefore we must not limit how God creates."

Larry Hench is a scientist for whom the Chautauqua idea is of paramount significance: all life is one and religion belongs everywhere. Good scientist that he is, he wants to know the truth of what he investigates. He is thus controlled, as scientist, by the scientific method, which is to pursue his research and let what he investigates disclose itself as it is, to reveal its own reality to him. He allows, in other words, the object of his investigation to control how he knows about that reality, how he formulates his knowledge about it, and how he proposes to verify it. As believer — as one for whom all life is one and for whom religion belongs everywhere — he takes another way of looking at the truth. Looking at, say, dead or inert bone as a scientist he will, by the forms of his own discipline, seek to exercise control over the object of his study.

But looking at the human beings who will benefit from his research, he will find another dimension: he now interacts with living beings. They, more than inert bone, have bearing on how he treats them, how they respond to him, and how he treats his discipline. Even more, then, when as believer he stands before the God he acknowledges as Creator, it is the one who is known, God, who controls how Larry Hench thinks about life, his science, his career, or his family. In both spheres, his science and his religious faith, he follows purposefully the true scientific method. He allows the reality towards which he turns to control the way by which he knows it: as scientist, concerned with the world of ceramics and tissue creation and regeneration; as believer, acknowledging that he can know God only as God can be known, i.e. through God's self-revelation and grace. Here is a scientist who therefore as scientist looks away from God to the nature of the world and a believer who as believer looks away from the world to the Creator.

In this regard, the book makes a significant contribution to the huge change that is going on in contemporary science in its understanding of the foundation of knowledge. Knowing what we know has become increasingly complex. The change in the contemporary discussions between the world of science and that of faith represents a marked departure from the dualism that bedevilled both science and religion from the time of the Enlightenment. It is a recognition that all life is one: a unitary and intelligible cosmos. It is a recognition that we need the work of both science *and* theology to help us understand that cosmos and learn what its ethical implications are in all the issues of life, whether at its beginning, in its multiple forms, or at its end.

The readers of this book will, I hope, come to its conclusion with three deepened convictions: first, of the need for scientists and theologians to be in conversation with one another about their core studies and reference points; second, of the need for the two disciplines to apply the same rigour to their pursuit of truth; and third, of the need to find an ethic that draws its integrity from what both scientist and theologian have both found in their respective disciplines.

Larry and June Hench and I share a love of music. It occurred to me once, listening to a symphony concert with them in the amphitheatre at Chautauqua, that by enjoying the beauty of the music we were also obeying God. The enjoyment was the obedience. As he has explained how he approaches his work as scientist, I have come to see that Larry Hench is a scientist who obeys God by the very rigour of his research and achievements. He has therefore helped to build a bridge between academic science and the practice of religious worship and prayer. This is the recovery of the spiritual perspective about which the founders of the Chautauqua Institution spoke. All life is one, and religion is everywhere.

I have pleasure, then, in commending the book *Science, Faith and Ethics* to readers.

Ross Mackenzie, Historian
The Chautauqua Institution
September 25, 2000

Preface

This book is an updated version of a series of five lectures I presented at the Chautauqua Institution Department of Religion seminar series, Chautauqua, New York, USA, July 28–August 3, 1991. The title of the lecture series was "The Impact of Science on Faith and Technology on Ethics". In these lectures I explored the various ethical issues associated with the growing use of human spare parts to prolong and enhance the quality of life. The lectures were based upon my 25 years of experience in developing new materials for use as medical implants. Of particular concern in the lectures was the problem of people expecting miracle cures when modern technology is used in medicine.

In 1996 I accepted the Chair of Professor of Ceramic Materials at Imperial College of Science, Technology and Medicine, University of London. I currently also serve as Director of the Imperial College Centre for Tissue Regeneration and Repair. I co-direct with Professor Julia Polak, Head of the Department of Histochemistry, a large multidisciplinary research programme in tissue engineering. Our objective is to activate the human body's repair processes to overcome the degradation and failure of ageing tissues. This research involves using the latest generation of science and technology for growing specialised cells in culture, including embryonic stem cells, and establishing a genetic basis for tissue regeneration and repair.

The need for new approaches to the repair of deteriorating tissues is expanding at an alarming rate as the world's population increases and the percentage of the population above the age of 60 increases even more rapidly. Nearly all options for repair or replacement of body parts have some degree of risk and lead to some degree of ethical concern. Alternative methods of repair, such as tissue engineering, that may involve genetic manipulation of cells, are especially sensitive. These concerns are addressed in this book.

An important goal of this book is to discuss the scientific and ethical issues associated with the beginning and ending of life and the preservation of the quality of life in between. A further goal is to examine the relevance of faith in an age of technological dominance. Throughout the book I explore the hypothesis that ethical dilemmas arise from a conflict of uncertainties. I examine the origin of these uncertainties and the importance of faith in dealing with them.

This book was started in the United States and was finished in the United Kingdom. Consequently the effects of science and the distribution of health care in both countries are discussed. The contrasts are intriguing. The US still has no national health care policy after 10 years of debate, whereas the national health care system in the UK is under increasing attack to become "more like the US system". In the book I examine some of the reasons for this curious state of affairs. I hope to have kept straight the differences and similarities between the two countries, but if I have erred I hope the reader will understand.

Larry L. Hench

1
Frontiers of Knowledge: Where Science Fiction Becomes Science Fact

We live in a world that worships knowledge.

Knowledge is power.
> — *James Murray, Editor*, Oxford English Dictionary, *in*
> The Surgeon of Crowthorne, *by Simon Winchester*

Those who know the least obey the best.
> — *George Farquhar*, Faber Book of Aphorisms

With knowledge comes truth.
> — *Anon. student graffito, Imperial College*

To know all is to be all.
> — *Anon. student graffito, Imperial College*

These and many similar sayings are hallmarks of our scientific age. A goal of this book is to explore two modern myths: that knowledge creates certainty, and that with knowledge comes security.

We begin by discussing the concept of the frontiers of knowledge and observe that there are limits. The objective of science is to probe these limits. However, just as a receding horizon can frighten a sailor, the uncertainty that results from exploring the boundaries of knowledge can be unsettling.

For many thousands of years the understanding of the physical world and man's relation to Nature was limited. It is conjectured that natural philosophers such as Sir Isaac Newton and Sir Francis Bacon comprehended almost the entirety of what was known about the natural world at their time. In contrast, at the beginning of this new millennium it is impossible for an individual to know even a small fraction of a single scientific discipline, such as chemistry, physics or biology. In my field, biomedical materials, there are several thousand scientific papers published each year. It is not possible to read every one; it is necessary to limit my reading to those that are of the closest interest to my field of research and to rely on reviews or book chapters for the other topics. A similar situation exists for every other field of science.

The New Oxford Dictionary of English defines knowledge as: "Facts, information, and skills acquired by a person through experience or education; the theoretical or practical understanding of a subject; the sum of all that is known. In philosophy, knowledge is a true, justified belief, certain understanding, as opposed to opinion."

Another definition is: "Familiarity, awareness or understanding gained through experience or study. Includes both empirical material and that derived by inference or interpretation."

Because of the overwhelming amount of information that has been generated by scientific inquiry during the last 300 years, it is useful to consider knowledge as three intersecting worlds.

The Three Worlds of Knowledge

The three worlds of knowledge are:

SELF, NATURE, SOCIAL

Knowledge of self involves knowing your physical, mental and psychological capabilities and limitations.

Knowledge of Nature encompasses all the so-called natural sciences; for example, physics, chemistry and biology.

Social knowledge involves understanding the consequences of the interaction of oneself with a second person, interactions with many people, or the interaction of oneself or of many people with nature; for example, fields such as sociology, anthropology, archaeology, economics and law.

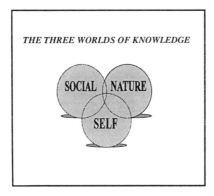

Figure 1.1 The three worlds of knowledge: nature, social and self.

The frontiers of knowledge exist at the boundaries of these three worlds.

Frontiers also are present where each world of knowledge intersects with another, i.e. self–nature, self–social, nature–social. The frontier where all three worlds of knowledge intersect, i.e. self–nature–social, is especially important (Figure 1.1). It is of special interest in this book because the personal, social, economic and ethical issues associated with the preservation and elongation of life involve all three worlds of knowledge.

The Limits of Knowledge

The belief that science creates certainty is a fallacy. This fallacy exists in the minds of many because they are unaware of or ignore the fact that there are limits to knowledge. The limits are present in three forms:

DISTANCE, TIME, THEORETICAL

We live in a world bound by space (distance) and time. This five-inch-long line of print takes about five seconds to read. With a finely calibrated scale and stopwatch you could establish with great accuracy the length of the line and the time of reading it. You could

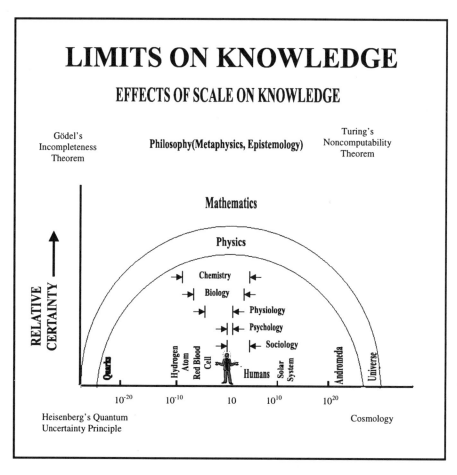

Figure 1.2 Limits on knowledge: effects of scale.

repeat the measurements many times and achieve a high level of certainty of both the distance and the time involved. However, if you attempt to measure something very small, such as the molecules that make up the print of the letters on this page, or something very long, such as the length of all the lines of print in this book, the errors of measurement are compounded and the level of certainty decreases. At the boundaries of measurement (i.e. the extremely small, such as the dimensions of the atoms within the ink molecules,

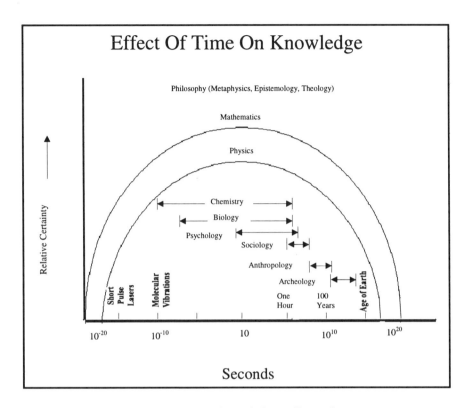

Figure 1.3 Limits on knowledge: effects of time.

or the extremely long, such as the distance from you to the outer edge of the universe) the level of certainty decreases immensely.

Figure 1.2 illustrates the effects of scale, or distance, on the relative certainty of knowledge. This is a hypothetical cross section of the sphere of knowledge with an axis of distance, in metres. At right angles one can draw another cross section of the sphere of knowledge with an axis of time and units of seconds (Figure 1.3). Man is assumed to be at centre of the world of knowledge (whether this assumption is justified has been debated for thousands of years). The relative certainty of measurements and observations of the natural world made on the scale of millimetres (10^{-3} m) to kilometres (10^{+3} m) is quite high, as illustrated in Figure 1.2. When the scale of an object, such as a red blood cell, is in the range of micrometres (10^{-6} m) the level of certainty of measurement and observation is still reasonably high, but is decreased to the limits of the optical microscope used.

Observation of the organelles inside a cell, such as the nucleus, or the chromosomes or strands of DNA inside the nucleus, involve dimensions of 10^{-8} to 10^{-9} metres and require the use of an electron microscope, where the wavelength of the electrons is on the scale of individual atoms. At this small scale the level of certainty is very much lower. To use the electron microscope the cell must be removed from its normal environment. This change of environment has a massive effect on the cell. It alters the morphology of the cell; most importantly, it kills the cell. The electron beam interacts with the molecules that make up the cell in order to produce an image. Electron microscope images provide important visualisation of the hierarchical arrangement of molecules that make up the structure of the cell (Figure 1.4). However, this knowledge of a cell is knowledge of a *dead* cell, not a living cell. There is a low level of certainty that information from a single electron microscope image is representative of living cells in living tissues in a living person. Only after repeated electron microscope observations of many cells combined with other types of analyses can the level of certainty be increased.

Observations of atoms that are at the scale of 10^{-10} metres and the sub-atomic particles that make up atoms, such as quarks (see Chapter 2), require the use of particle beams that are very much higher in energy than the electron beams in an electron microscope. The smaller the particle, the higher the energy required to observe it and the lower the certainty of the observation. Thus, at the limit of knowledge in the range of 10^{-20} metres, the level of certainty becomes very low indeed, as illustrated in Figure 1.2.

Likewise, as observations are made at scales that become very much greater than the dimensions of a human, such as the distance to the nearest galaxy, the level of certainty also decreases. When the scale of interest is the size of the universe, which we discuss in Chapter 2, the level of certainty is almost infinitesimal.

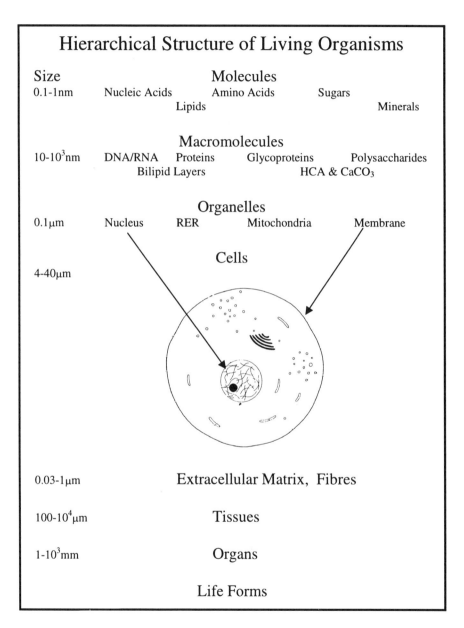

Figure 1.4 Hierarchical structure of living organisms.

Figure 1.2, showing the effect of scale on certainty of knowledge, indicates that the level of certainty also depends upon the *field* of knowledge. It suggests that the level of certainty of the knowledge of physics is higher than that of chemistry, which is greater than biology and physiology, and so forth. This controversial perspective is based upon the conclusion that the level of certainty is inversely proportional to the number of items being observed. The level of certainty in the physics and chemistry of hydrogen atoms, which have only one electron in orbit around one proton in the nucleus, is very high. Combining two hydrogen atoms with one oxygen atom to make water creates a molecule that is critical to life. It also creates a molecule that has enormous variability in its physical and chemical behaviour, depending upon its temperature, its pressure and the number of neighbouring molecules. The uncertainty in biological systems is high. This is because there are generally many more atoms in biological molecules undergoing interactions than are present in well-controlled chemistry experiments, for example. As the number of particles becomes very large, such as the number of water molecules in the atmosphere, their interaction becomes chaotic and predictability becomes impossible. Certainty of knowledge of the behaviour of very large systems is limited. Thermodynamic quantities can be defined which are certain although the behaviour of individual atoms is not. Chaos theory shows that patterns of behaviour of large systems are repeatable but not predictable, i.e. they are uncertain.

We now know that the physical knowledge about even a single particle of matter, such as an electron or a neutron, has a theoretical limit. Heisenberg's uncertainty principle is part of the foundation of quantum mechanics, which provides the theoretical basis of much of modern physics, chemistry and biology. The uncertainty principle states that one cannot simultaneously define the momentum (mass times velocity) *and* the location of a particle. It also states that one cannot simultaneously establish both the energy *and* the time of a particle. The origin of the uncertainty is a consequence of all the components of matter having a dual wave–particle nature. Electrons and neutrons and even atoms sometimes behave as waves in their interaction with each other. Under different circumstances they behave as if they were solid particles. Thus, there is a limit to the certainty or predictability of their interaction. The consequences of these uncertainty relationships are important. If a particle moves slowly, its momentum is small. Therefore its wave function extends over a considerable space, so that the particle can interact with anything in the region. A slow-moving neutron acts as if it were larger than a fast neutron. Control of nuclear power reactors depends upon this dual nature of matter. Without careful continuous control the uncertainty inherent in Nature leads to disasters such as Chernobyl and unpredictable, chaotic consequences.

Heisenberg's uncertainty principle shows that there is a limit to knowledge at the scales of the very small. Our inability to provide the necessary input for Einstein's theory of relativity imposes a limit on knowledge at the scales of the very large. The sphere of knowledge illustrated in Figure 1.2 indicates that there are at least two other limits to our ability to understand and predict the world around us. These are theoretical limits of knowledge. The limits are independent of observation. They are known as Gödel's incompleteness theorem and Turing's noncomputability theorem.

Gödel's theorems, published in 1931, state, in paraphrase, "You may know it but you can't prove it."

As discussed in the *Cambridge Dictionary of Philosophy*, "Gödel's First Theorem is a disappointment to any theory constructor who wants his theory to tell the whole truth about a subject."

Thus, there is a theoretical limit on logical analyses, i.e. a fundamental limit on certainty.

The Gödel theorems were extended in 1936 by A.N. Turing in his theoretical analysis of computability on a Turing machine.

Turing's noncomputability theorem states, in paraphrase, "You can't prove it by computing it."

As stated in the *Cambridge Dictionary of Philosophy*, "There is no hope of proving any of the Turing theses, for such proof would require a definition of "computable" — a definition that would simply be a further item in the list, the subject of a further thesis. Turing proved the unsolvability of the decision problem for logic."

The study of the nature of knowledge, called epistemology, also indicates that there is a theoretical limit on certainty of what we know or can know. The problem lies in defining the criterion for judging the truth or falseness of the appearance of things.

> "The critical question is, How can we specify what we know, and how can we specify how we know without specifying what we know? Is there any way out of this circle? A cogent epistemology must provide a defensible solution to it. Contemporary epistemology still lacks a widely accepted reply to this urgent problem."
>
> — *Cambridge Dictionary of Philosophy*

Let us return to Figure 1.3, which shows the effect of time on the certainty of knowledge.

The immense success of Stephen Hawking's book, *A Brief History of Time*, published in 1990, shows the intense interest of many people in the subject of time. Or, it at least gives the impression of such an interest. This interest has continued to the present.

For example, in a recent issue of *The New Scientist* (21 August 1999, pp. 46–47): "Marcus Chown invites you on the ultimate trip through the physics of eternity, time after time." His exciting review of *The Five Ages of the Universe*, by Fred Adams and Greg Laughlin, concludes, "The sheer chutzpah of physicists is amazing. Not content to speculate only on the first 10^{-43} seconds of the Universe, they believe they can map out at least roughly the next 10^{+100} years and beyond. But the point is not that we are sure the future will turn out the way Adams and Laughlin say. Rather, it's that in 1999, for the first time, we have in our hands a self-consistent model of physics and cosmology."

Chown continues with a caution, "Of course self-consistency does not necessarily imply correctness. Adams and Laughlin have cranked the shiny new machinery of physics. They have taken us on the ultimate journey through space and time, probing for — and finding — the limits of our understanding."

When the duration of time is in the scale of our normal perception — seconds, minutes and hours — our level of certainty of observation and knowledge is high, as illustrated in Figure 1.3. But, when we extrapolate backwards in time — the domain of historians, archaeologists, palaeontologists, geologists, etc. — the level of certainty of knowledge decreases with the number of years of extrapolation.

Similar extrapolations forwards in time, such as those cited above, suffer even more in loss of certainty: this is because, by definition, we cannot know *a priori* the unpredictable, i.e. natural disasters, such as volcanic eruptions, or collisions with meteorites and human developments, such as new scientific discoveries or cataclysmic wars.

An intriguing aspect of time is our ability to extrapolate backwards and forwards. Although we only experience time moving in one direction, as discussed in Hawking's book, cosmologists freely project physical events in time backwards by 10^{17} seconds and forwards by equally large increments. The level of certainty of these extrapolations is extremely low, as depicted in Figure 1.3. On an almost weekly basis scientific articles, and even books, appear that modify values of the age of the universe, the time when life appeared on earth, and the time when *Homo sapiens* evolved from ape-like predecessors. These topics are discussed in more detail in Chapters 2 and 3.

The uncertainty in time has recently received an even more shocking jolt than presented by Hawking. Another British physicist, Julian Barbour, proposes in his book *The End of Time* that time does not even exist. He believes that time does not exist as a fundamental property of the universe. He proposes that we experience only transitory moments called "nows". Although our brains essentially integrate the instantaneous "nows" into what we think is a continuous flow of "time", it is in reality only an illusion. This controversial perspective of time bears a haunting similarity to the uncertainty principle, to Gödel and Turing's theorems, and to the epistemological limit on certainty discussed earlier.

Final thoughts on the frontiers or limits of knowledge can be drawn from Peter Medewar's insightful book *The Limits of Science*. He reminds us that Bacon in 1623 spoke of three limitations on comprehending the universal nature of things. First, do not put our trust in knowledge such that we forget our mortality. Second, we must apply knowledge to achieve that which is good rather than evil. Third, do not presume to attain the mysteries of God by studying Nature.

This first limit on knowledge is expressed by Medewar to be "The existence of questions that science cannot answer and no conceivable advance of science would empower to answer." He lists three such unanswerable questions:

> How did everything begin?
> What are we all here for?
> What is the point of living?

Some readers will take the position that such ultimate questions are valid only for philosophers, that they are questions of metaphysics rather than physics. Other readers may debate that these questions are central to our existence and that science has provided the answer to at least the first question and a means of examining the other two.

I submit that the frontier, or boundary, between science and faith lies somewhere between these questions. We must explore this boundary in order to live a meaningful life. Our objective in the following chapters is to make this exploration.

Bibliography

1. *Modern Cosmology and Philosophy*, John Leslie (ed.), Prometheus Books, Amherst, New York, 1998
2. *Mathematics: The Loss of Certainty*, M. Kline, Oxford University Press, Oxford, England, 1980
3. *The Emperor's New Mind*, R. Penrose, Oxford University Press, Oxford, England, 1989

4. *The Limits of Science*, Peter Medewar, Oxford University Press, Oxford, England, 1988
5. *The Cambridge Dictionary of Philosophy*, Robert Audi (ed.), Cambridge University Press, Cambridge, England, 1995
6. *The Five Ages of the Universe*, Fred Adams and Greg Laughlin, The Free Press, London, 1999
7. *A Brief History of Time*, Stephen W. Hawking, Bantam Books, New York, 1989
8. *The End of Time*, Julian Barbour, Weidenfield & Nicholson, London, 1999
9. *The Edges of Science: Crossing the Boundary from Physics to Metaphysics*, Richard Morris, Prentice-Hall Press, New York, 1990
10. *About Time: Einstein's Unfinished Revolution*, Paul Davies, Penguin Books, London, England, 1995
11. *Three Scientists and Their Gods: Looking for Meaning in an Age of Information*, Times Books, New York, 1988
12. *Who's Afraid of Schrödinger's Cat? The New Science Revealed: Quantum Theory, Relativity, Chaos and the New Cosmology*, Ann Marshall and Danah Zohar, Bloomsbury Publishing, London, 1997
13. *The Second Culture: British Science in Crisis: The Scientists Speak Out*, Clive Cavendish Rassam, Aurum Press, London, 1993
14. *Hidden Histories of Science*, Robert B. Silvers, The New York Review of Books, New York, 1995
15. *The Tao of Physics*, Fritjof Capra, Shambhala Publications, Colorado, 1975
16. *The End of Science*, John Horgan, Broadway Books, Bantam Doubleday Dell Publishing Group, Broadway, New York, 1997
17. *Modern Cosmology and Philosophy*, John Leslie (ed.), Prometheus Books, Amherst, New York, 1998

2
Origins of the Universe and Matter

Introduction: A Conflict of Uncertainties

For thousands of years mankind's ethics were based upon the pragmatic observation that "might makes right". As societies evolved, the concept that there existed a source of "right" beyond humans began to be established. Whatever was observed to be "good" and "right" was attributed to a myriad of supernatural beings or "gods". Whatever was "non-good" or "evil" was also attributed to a host of equally powerful gods or demons.

In order to ensure good and thwart evil, this primitive theology required making sacrifice and paying homage to the various gods.

It was not necessarily a very happy theology, but it was a theology that offered a great deal of certainty. When things went well, you had kept the god(s) happy and were rewarded. When things did not go well, you had failed to keep the god(s) happy and were punished. Consequently, even primitive religions provided an image of certainty in a world of uncertainty. Most people accepted the word of religious leaders and seldom questioned "higher authority", even if it was unsatisfying.

Humans had little capacity to know their primitive gods, who took no personal care of mankind. The whimsical, unpredictable character of such gods was accepted as part of the unpredictable character of the natural world.

In the midst of a pantheistic world of multi-gods of good and evil arose the lonely voice of monotheism of the Jewish patriarchs. Their lofty vision of an all-powerful, all-knowing, all-caring God resulted in a unique theology. Nearly all the western world are the inheritors of this theology, which says that God, the creator, knows and cares about man, the created.

14

Judaic theology also asserts that "Humans can know God" by means of God's Word revealed as Law. Thus, the relationship between God and humans is through the Revealed Word and the Law contained within the Word.

This remarkable concept greatly increased the level of certainty in the Hebrew religion.

The source of good is God. God gives the Law so humans can know God. To know and follow the Law is to know and follow God and therefore should yield good. Not to know or not to follow the Law is not to know or not to follow God and therefore should yield non-good or evil.

This theology yields the perceived certainty that if good befalls a person it is because he or she has followed the Law. Conversely, if evil comes it is because he or she has not followed the Law, i.e. has sinned. Thus, the greater the troubles, the greater must be the sin.

Most of the time this relationship of God → Law → human is followed by good. There is a fairly high level of certainty and confidence in such a religion. However, we all know that good does not always befall those who follow God's Law. This is a fundamental problem and creates uncertainty in Judaic–Christian theology. God's answer, "Trust me", to Job's questions of "Why me, Lord? What did I do to deserve this?" is no more satisfactory now than it was 3,000 or so years ago.

However, for thousands of years theological uncertainties such as those debated in the book of Job had little impact on ethical issues for people since there was little choice but to accept natural disease, death and disasters. Prayer sometimes helped and sometimes it didn't. So, you had no alternative but to accept.

This acceptance of natural uncertainties in life and death has now changed. During our lifetime the image of certainty and the source of answers have shifted from religion to science.

The media trumpets that even the two great questions, i.e. the origin of the universe and the origin of life, are answered by science, or nearly so. The few uncertainties that are left, such as events of the first millionths of a second of the big bang or the complete encoding of the human genome, are regarded as details to be resolved within a few years and a few billion dollars.

Likewise, rapid human acceptance of the calculational superiority of computers and the advent of so-called artificial intelligence add to the impression that science and technology have the answers and certainties. Religion is left only with questions and uncertainties.

Ethical dilemmas arise, however, because science and technology offer only partial answers to most of the important questions. The interpretation of the partial answer still rests on the individual and the family. The uncertainties arising from partial answers are often more distressing and unsatisfying than no answer, as Adam and Eve were warned. Faith becomes a fall-back position. However, faith is now eroded by debate and dissension.

It is important to explore the degree of uncertainty in the scientific understanding of the origin of the universe, the origin of life, and the evolution of life. We find that the answers are not always convincing. We also find that, in spite of the uncertainty, there is a comfortable degree of harmony of these scientific viewpoints with the descriptions presented in the Revealed Word — the Bible.

In later chapters we examine the complex ethical issues facing the health care and biotechnology fields today and tomorrow. These ethical issues will be viewed from our central position of a conflict of uncertainties. We will argue that personal faith in a loving and caring God is essential in resolving these conflicts.

Stories of Creation

"Genesis" 1 describes the beginning as most of us were told:

> "In the beginning God created the heaven and the earth. And the earth was without form, and void; and darkness was upon the face of the deep. And the Spirit of God moved upon the face of the waters. And God said, Let there be light and there was light. And God saw the light, that it was good: and God divided the light from the

darkness. And God called the light Day, and the Darkness He called Night. And the evening and the morning were the first day. And God said, let there be a firmament in the midst of the waters and let it divide the waters from the waters. And God made the firmament, and divided the waters which were under the firmament from the waters which were above the firmament: and it was so. And God called the firmament Heaven. And the evening and the morning were the second day. And God said, Let the waters under the heaven be gathered together unto one place, and let the dry land appear: and it was so. And God called the dry land Earth; and the gathering together of the waters called He Seas; and God saw that it was good. And God said, Let the earth bring forth grass, the herb yielding seed and the fruit tree yielding fruit after his kind, whose seed is in itself, upon the earth: and it was so. And the earth brought forth grass, and herb yielding seed after his kind, and the tree yielding fruit, whose seed was in itself, after his kind: and God saw that it was good. And the evening and the morning were the third day. And God said, Let there be lights in the firmament of the heaven to divide the day from the night; and let them be for signs, and for seasons, and for days, and years: And let them be for lights in the firmament of the heaven to give light upon the earth: and it was so. And God made two great lights; the greater light to rule the day, and the lesser light to rule the night: He made the stars also. And God set them in the firmament of the heaven to give light upon the earth. And to rule over the day and over the night, and to divide the light from the darkness: and God saw that it was good. And the evening and the morning were the fourth day."

We return to "Genesis" for the fifth, sixth and seventh days in Chapter 3.

For now, let's compare the beautiful words of "Genesis" 1, verses
1–19 with another story of the creation also originating many
thousands of years ago, the story of Brahma from the Hindu religion.

> "Before the sky or stars were made there was darkness.
> The darkness was everywhere, and it was empty. The
> darkness was endlessly rippling and eddying throughout
> the universe. Its ripples gradually transformed themselves
> into sound. A word began. At first the word was no
> more than a whisper, but the word swelled and grew
> to a billow of a sound, endlessly repeating, folding back
> on itself, coiling and twisting till it filled all space.
> Om…Om…Om…Om… .
>
> "As the word unfolded and spread, calm as a heartbeat,
> it turned the rippling universe into an endless, unfathom-
> able ocean. Deep in the water bobbed a seed. As the
> ocean currents ebbed and flowed they carried the seed
> to the surface. It became a golden egg. The egg rocked
> on the water and the sacred word "Om" went on cradling
> it. The sound was in and out and round about and inside
> the eggshell. As a long time passed it formed itself into
> Brahma the First Father, Creator of Worlds.
>
> "Brahma hatched like a chick from the golden egg.
> He made the sky from half of the shell and the earth
> from the other half. He set air between them to keep
> them apart. The earth shell bobbed on the sea until
> Brahma anchored it with rocks and mountain peaks.
> When the earth was ready Brahma drew out of himself
> six elements: thought, hearing, sight, touch, taste and
> smell. He blended the elements to make living things of
> every kind. One thing he kept from living things: thought.
> Brahma kept thought locked into himself.
>
> "After a while Brahma divided himself and made
> Sarasvati. He fell in love with her and after a wedding
> night of 100 years in a secret cave, Manu, the first human

being, was born. Brahma gave Manu eight gifts: the five senses, the power of movement and reproduction, and the greatest gift of all, the power of human thought."

There is a beautiful similarity of the creation myth of Brahma to that of Ra, the sun god of Egypt.

"In the sea that was there before the world, a single, perfect flower appeared. It was a lotus. It had a green stem, flat waxy leaves, and milk-white petals folded like a fist. The lotus bud bobbed on the water in the darkness, and inside safe as a child in its mother's womb slept the baby sun. The sun's name was Ra, and he and the lotus were the only living things. As the lotus bud fattened, so baby Ra inside it slept and grew waiting to be born. The lotus that was Ra's floated in the darkness before time began. One by one its petals uncurled and spread until it lay fully open, bobbing on the sea like a saucer. As the lotus opened it filled the darkness with the first and the sweetest flower scent the universe has ever known. The scent tickled baby Ra's nostrils and woke him up.

"Ra stretched his arms and legs and opened his eyes — and at once dazzling beauty streamed out of them and sent the sea spray skittering. In the instant Ra was born he grew to full maturity. One second there was a chubby baby gurgling in a lotus. The next instant there was a grown prince towering from horizon to horizon. He imagined children and grandchildren and they too instantly appeared; the air and rain, the sky above and the sea and earth scatter across it from side to side.

"Ra looked down at his new creation and the dazzle of his own sunbeams glittered back at him from a thousand waves. The glare made his eyes water. The water drops fell to earth and turned into insects, fish, birds and human beings. So, the world's creatures were

> born from tears, and tears and sorrows have been their
> nature ever since."

Nearly all cultures have creation stories and most have a common theme. Out of darkness came light. Out of nothingness came something. Matter was born from non-matter. Matter was then divided and separated into earth and sky and all things living and non-living.

The biblical creation story has some of the most beautiful verses in literature. It also has some of the most controversial. The basis for many of the perceived conflicts between science and religion is the literal interpretations of these 12 verses and these 6 days of creation. There are millions of devout believers, Jewish, Christian and Muslim, who accept the literal interpretation of these words.

Conversely, most scientists believe that the universe was created some 7 billion to 20 billion years ago, with the most recent and best measurements within the range of 10 billion to 15 billion years. Estimates of the ages of stars in globular clusters fall within the range of 11 billion to 16 billion years and the age of our galaxy is between 9 billion and 16 billion years. Most scientists accept that the earth's age is approximately 4.5 billion years. They also generally agree that the first, living, one-celled organisms appeared about 3.5 billion years ago, and modern man, *Homo sapiens*, finally evolved from a long line of progressively more complicated life forms about 1–3 million years ago.

The discrepancy between 6 days of creation and 16 billion years is pretty hard to ignore. Fortunately, for most of us we can just shrug and say, "So what?" However, this difference affects the attitudes of many people. As recently as 1981 the Arkansas state legislature in the United States tried unsuccessfully to require teachers to spend equal time teaching scientific creationism, the literal interpretation of biblical creation, and teaching evolutionary creation science.

A federal judge ruled the state law as unconstitutional due to a conflict of church and state. His ruling said, "There wasn't any science in scientific creationism." A solution for many fundamentalist churches is to operate their own schools and teach their own curricula, which can give equal time to the two perspectives. In 1999 the School Board of the state of Kansas eliminated the teaching of evolution in state schools as a requirement for obtaining a high school diploma. Chapter 4 discusses this controversy.

Thus, the scientific and religious issues associated with the origin of the universe can be highly relevant on a personal level, especially if you hold views opposite to those of your spouse, family, neighbours, boss, colleagues, local school, church or car mechanic.

A goal of this chapter is to examine how much fact and how much fantasy there is in the scientific understanding of the origin of the universe. We will explore what is known and agreed on by the majority of astrophysicists and cosmologists. We will also try to establish what remains uncertain and controversial.

In other words, our goal is to ask whether present day science on origins of the universe is based on fact. Or has a scientific myth arisen to replace the religious myths? Also, how do these scientific facts compare with the story of creation as revealed in the Bible?

The Big Bang

The most generally accepted scientific theory of the origin of the universe is called the "big bang". The big bang is often referred to as "the standard model". The first point to recognise is that the big bang theory is not just one theory; it is a family of theories. As in any family, some members are easier to accept wholeheartedly than others. And, also, just as in any family, the consequences of accepting each member are usually quite different.

For example, one of the big bang models results in a universe that is 20 billion years old. Another model results in a universe that is only 7 billion years old. The cosmologists' challenge is to create a self-consistent theory that eliminates this difference.

Before discussing the differences within the big bang theory, let us first consider the scientific facts that led to this theory.

In 1929 the American astronomer Edwin Hubble discovered that the universe was in a state of rapid expansion. Based on spectroscopic measurements, he showed that all the galaxies that can be seen were rushing rapidly away from each other. He found that the further a galaxy is from earth, the faster it is moving away from us.

The discovery of the galaxies flying away from each other today implies that there was a time when they were closer together, even a time when they did not yet exist as galaxies but instead existed as matter compressed together in a very dense mass. The big bang occurred when this very dense matter rapidly expanded, resulting in a rarefied, almost uniform distribution of matter, similar to a fog, mist or cloud of dust. Stars and galaxies formed when matter was attracted by gravity to condense together in regions within this dust cloud.

Using the outward speed of the galaxies, it is possible to assign a date to the origin of the universe. To do so, cosmologists assume that the expansion is slowed by the gravitational force that acts between all particles of matter, just as a ball that is thrown up in the air is slowed in its ascent by the gravitational attraction of earth.

Since we know this gravitational force, from the upward speed of the ball we can calculate where it was at any earlier time and, in particular, we can calculate when it left the hand that threw it. Similarly, knowing the present speed with which galaxies are flying apart, and knowing the gravitational forces that act to slow them down, cosmologists can calculate the time at which all galaxies would have been at zero distance from each other. This time defines the beginning of the universe, i.e. the big bang.

This is not an easy calculation, as you might surmise. Even with more than 50 years of data collection and theoretical calculations, astronomers and cosmologists do not agree on the rate of expansion and do not agree on the age of the universe. One set of assumptions yields 7 billion years. Another set yields 20 billion years. The range quoted by the US National Academy of Sciences as the best estimate is 10–15 billion years. The spread in calculated values is due to a combination of uncertainties in several factors: (1) the rate of expansion, (2) the amount of mass in the universe, and (3) the type of mass in the universe.

There are several big problems with the big bang calculation. The first problem is that the calculation requires knowing the distance to galaxies and also how fast they are moving apart. Speed is measured quite accurately because we have very good benchmarks; the chemical elements that make up the stars, such as helium or hydrogen, emit light in the same manner as the same elements do here on earth when they are heated. Astronomers can measure very accurately the shift in the spectra of these elements emitted from the stars (called the redshift) and tell how fast they are moving. However, distance measurements are very uncertain except for the nearest galaxies. It is the galaxies that are very far away that provide the distances needed to calculate the age of the universe. The uncertainties in calculating distances give rise to part of the 7–20-billion-year variation in time estimates for the big bang.

Another big problem in the big bang calculation is knowing how much the galaxies have slowed down from the start of the big bang and therefore how much the expansion of the universe has slowed. The reason this is a problem is that it depends on the amount of mass in the universe and the "type" of mass. Some models require invisible, so-called "dark matter" to be present.

Cosmologists know that gravity should act as a brake on the movement of neighbouring objects, like galaxies, and therefore they must slow down, as the distance between them increases.

The density of mass in the universe determines how much the galaxies slow down. If the amount of matter per volume in the universe is less than a critical amount, which is about three hydrogen atoms/cubic metre, there will never be enough slowing and the universe will expand forever. This is called an "open universe". An open universe is generally infinite in nature.

If the density of matter in the universe is greater than this critical amount, the speed of separation will get slower and slower until gravity stops the expansion. If this happens the galaxies will all start to collapse back upon each other, leading to what is call the "big crunch". This model is called a "closed universe model". An important feature of a closed universe is that it is finite. It can also result in oscillations that go on in cycles forever, and, therefore, does not require a beginning.

Of course, it is also possible that the amount of matter in the universe is just equal to this critical amount. If that is the case, the universe is neither open nor closed; it is flat. A flat universe is also infinite, but it differs from an open universe in that the expansion slows progressively down essentially to zero, though it never stops.

What is the evidence for these alternatives: a closed, flat or open universe? Estimates of density of matter in the universe indicate that there is presently just less than the critical amount and close to the critical density. That generally rules out a closed universe and an oscillating universe and results in a Creator and a starting creating instant of the big bang.

However, the closeness of the measured density to the critical value gives rise to another problem: the flatness problem. The observed fact that we are very close to the critical density today, by a factor of 2–10, means that very early in the big bang, when all the particles were very close together, the universe must have been extremely, extremely close to a critical value. If the density value were smaller by even a minuscule amount, everything would have flown apart too rapidly for atoms to form, such as those that make up the earth and us. If the density value were ever so slightly larger, the particles would have all collapsed together again converting back into energy.

These theories are all a consequence of Einstein's theory of relativity, which also produced the famous equation $E = mc^2$. This equation says that matter (m) can be converted to energy (E). We are all too familiar with one consequence of that conversion by seeing, in films and on TV, the mushroom cloud of the atom bomb. The enormous destructive power of an atom bomb is due to uranium atoms being split into two atoms of just a little less mass. The difference in mass is released as energy. We are also all familiar with the concept of the hydrogen bomb, which is based on hydrogen atoms being forced together to form helium atoms. This releases even more energy because a helium atom has less mass than the hydrogen atoms. The fusion reaction in the hydrogen bomb is the same as in the sun. So, the conversion of matter to energy is easily accepted as science. It is experimentally tested, and observed, and verified, and thereby meets the generally accepted criteria for science — prediction, repetition and verification.

The equation $E = mc^2$ also predicts that energy can be changed into matter. That concept is much harder to perceive and understand. Exploring the conversion of energy into matter requires cyclotrons and synchrotons, the tools of high-energy physicists, and is not easily observed in television documentaries. The transformation of energy into matter is also what is believed to have happened in the big bang, thereby creating the universe.

The first forms of matter to appear from the big bang were the fundamental particles, now called quarks and leptons by high-energy physicists. The quarks make up protons and neutrons, which are the basis of all matter, the stars, the planets, and us. Quarks have a mass and gravitational attraction. Quarks as particles always occur together in groups of three, which is behind the origin of the name, coined in 1964 by the physicist Murray Gell-Mann, who took the term from James Joyce's novel *Finnegans Wake*: "Three More Quarks for Muster Mark".

You may ask, "What was present *before* the particles?" The answer agreed on by most physicists is the same answer you get when you read "Genesis" and many of the creation myths — before the particles with mass were particles without mass, called photons — or, in everyday language, "light". In "Genesis" 1 God says, "Let there be light", and "there was light", and "it was good".

Light is energy. The amount of energy in light can vary enormously. The particles that constitute this energy, the particles that make up light, are called photons. The warmth of the sun on your face is energy. It is due to low-energy photons called infrared light. You can feel these low-energy photons by means of heat sensors in your skin. But you cannot see infrared photons. The light by which we see is due to photons with frequencies in the visible range of the electromagnetic spectrum. Visible photons contain 1000 times more energy than infrared photons. That is why a stove top that is just warm does not burn you. It emits infrared photons, which are low-energy. However, a burner that is glowing very red will burn you badly. It has very much more energy.

Thus, receptors in our skin can feel the presence of low-energy radiation, i.e. heat, or the absence of it, i.e. cold, but only over a very limited range of energy. When the energy becomes very low, less than the infrared, we cannot detect it with any of our senses. When the energy becomes very great we also cannot see, feel or hear it. The receptors in our eyes are sensitive to only a very limited range of energy. Photons that we cannot see we no longer call light. We give them names like X-rays or gamma rays or ultraviolet rays. Most of us have had X-ray examinations and know that the particles go right through us. The X-ray diagnosis is based on the fact that X-rays pass through us with different intensities, depending upon whether they collide with bone, which is dense, or with soft tissues, such as muscle or skin, which are low density. Gamma rays or neutrinos go through us, presumably without interacting with matter. The low end of the energy spectrum, electromagnetic radiation, such as radio waves, has so little energy that we do not know that the waves are present until we convert them into sound waves in a radio, telephone or TV.

So, let us return to the story of creation as described by the big bang. At the beginning there was only energy, energy which we could not perceive, even if we were there as an observer, just as we cannot see or feel X-rays. The energy was millions and millions of times greater than X-rays. The energy was so large that the numbers are not meaningful to the human brain. Then, in a process that physicists only speculate upon, the energy began to transform. Symmetry between forces became broken. Particles emerged from the energy: electrons appeared, and quarks were created which combined with the strong force transmitted through gluons to form protons and neutrons, which in turn combined with the electrons to form matter as we now know it.

The description of this sequence of physical events in the early seconds of the creation of the universe has a familiar sound to it, at least as written by Richard Morris in his elegant book *The Edges of Science*:

> "Before that time, referring to the Big Bang fireball of 16 billion years ago, the universe was filled with photons and free electrons that were moving much too rapidly to be captured by nuclei and form atoms. These electrons interacted with any light that came their way and absorbed, scattered, and re-emitted it in various different directions. The effect of all this was to produce a kind of cosmic fog. If there had been any conscious observers at this time, they would have found the universe to be nearly opaque, although filled with a brilliant glow. Then, as the universe expanded in the Big Bang, its temperature dropped. As the universe cooled, the electrons gave up some of their excess energy and began to form atoms. As they did, the fog began to lift."

In just one minute, as we now perceive time, the hydrogen and helium that make up most of the universe came into existence.
I repeat "Genesis" 1–3:

> "In the beginning God created the heaven and the earth. And the earth was without form, and void; and darkness was upon the face of the deep. And the Spirit of God moved upon the face of the waters. And God said, Let there be light: and there was light. And God divided the light from the darkness."

The scientific and biblical versions sound similar. Gerald Schroeder, in his intriguing book *Genesis and the Big Bang*, discusses the subtle distinction in the Hebrew words in "Genesis" 1.1–1.3 translated into English as "water" and "light". He shows that the ancient words were slightly different in meaning from the "water" of the seas or the "light" of the sun. They were words of much broader meaning.

Thus, translations of the ancient Hebrew by Maimonedes (1135–1204) and Nahmanides (1194–1270) over 700 years ago indicate that these subtle distinctions that are present in the divine revelation of the Bible result in considerable consonance with the physical distinctions described as the first minutes of creation by the big bang theory.

If you seek consonance between the big bang and biblical creation, in spite of the uncertainties, Gerald Schroeder provides a starting point. The big discord to resolve, of course, is the difference between 6 days of creation and 16 billion years. Schroeder reminds us that in addition to the energy–mass relationship, Einstein's theory of relativity predicts that time slows down as the speed of an object increases. When astronauts return from a space shuttle trip, they are a few moments younger than their spouses. For massive particles travelling near the speed of light, time essentially stops.

Consequently, the theory of relativity indicates that at the beginning, when the universe was without form, time was compressed or did not exist as we now perceive it. One day for God could be the same as several billion years for us. Only after the universe cooled down and matter formed, did time, as we experience it, begin to have meaning within the human context.

Thus, Schroeder argues it is partially the limitations imposed by our human perception of time that cause the inconsistency between divine revelation in the Bible and cosmological science. He suggests that it is reality that the actual time of creation, as perceived and measured by humans today in a cold universe, did require 7–20 billion years in human time. However, it may also be equally true that the actual time of creation, in "God's time", was only a few days. Thus, there is no inconsistency between the revealed Word and the scientific word. There is also at present no scientific principle that can be used to resolve this difference between cosmology and theology.

The concept of relativity of time is difficult to grasp or accept. However, clocks in orbiting spacecraft moving at 24,000 mph have been proven to slow down. That is fact. If a manned spacecraft could leave earth and travel to the stars at near the speed of light, it would take hundreds of earth years to make the round trip. When the astronauts got home, tens of generations of their families would have lived and died, but they would have aged only a few years. This is the stuff of science fiction. But, it is based on science fact, as we know it today.

Science or Faith?

Thus, the theory of relativity provides a scientific base for describing the origin of the universe, the creation of matter from energy, and the birth of time. However, does the theory of relativity provide a basis of certainty for us? Is it a substitute for faith? My personal opinion is that the answer is no. There are too many problems and too many alternative theories for us to accept present day cosmology as an alternative to faith in a Creator and a Revealed Word.

Instead, it seems essential to accept that the finiteness of human senses and our restricted perception of time and reality will always impose a limitation on the interpretation of scientific theories. Although we can measure many of the interactions between matter and energy, our understanding of these interactions must be limited by our measurement techniques and by our human finiteness in interpreting the data. In modern physics it is nearly universally accepted that there is an inherent uncertainty in the measurement of all particles that make up matter.

The Uncertainty Principle

This uncertainty, called the Heisenberg uncertainty principle after the pioneering quantum physicist Werner Heisenberg, is due to the fact that all particles have both a particle-like and a wave-like character. Consequently, protons, neutrons, electrons and atoms have a fuzzy characteristic like a ball of cotton, or like ripples in a pond. Their fuzzy character makes it impossible by this fundamental principle, in all practical situations, to detect precisely where the particle is or to know precisely its energy. The uncertainly is part of being real. It is part of existence. It cannot be explained away. It has to be accepted; and it is accepted as science.

These concepts are used to explain the properties of matter in the field of physics called quantum mechanics. Uncertainty is a fundamental component of quantum mechanics. A second fundamental component in the theory is the fact that each particle is discrete and individual, with its own characteristic energy and wave function. Boundary conditions imposed on particles, such as the number of neighbouring particles and their energy of interaction, affect both the energy and the fundamental character of the particles. If you change the boundary conditions you change the nature of the particle and you can never know for sure the details of the change. The uncertainty principle says we must be satisfied with knowing only the *probability* of how the particle behaves in the real world.

These concepts underlying quantum mechanics and the limitations of our observations of the physical world are similar to our limitations of relating *to* the physical world. The uncertainty principle says there is a limit to what we can know about matter and nature. As Paul Davies summarises in his book *About Time*, the uncertainty principle also limits what we can know about time. He states, "The core difficulty with quantum time harks back to the very notion of Einstein's time: there is no absolute and universal time. My time and your time are likely to be different, and neither is 'right' or 'wrong'; they are equally acceptable."

The Search for Unification

At the farthest boundaries of our sphere of knowledge, depicted in Figures 1.2 and 1.3, the very small and the very large converge with times that are very short and very long. Understanding this convergence is the great challenge of physics. This challenge has a name. In Einstein's era it was called the grand unified theory. Today it is called the theory of everything.

Einstein failed in his quest for unification. The conundrum he faced was the following: as cosmologists' projections move backwards in time and get ever closer to the origin of the universe, 10^{-43} seconds from the big bang, all the mass of the universe is drawn ever closer together, converging towards a single particle which was created from pure energy. Under these forbidding conditions quantum effects, i.e. quantum gravity, must be reconciled with relativistic effects. This is a formidable obstacle that baffled even the genius of Einstein.

Theorists today propose a solution to this problem. It is an elegant solution with roots in pure and applied mathematics. They assume that the fundamental particles and forces of nature are not described mathematically as points but instead are represented as strings. It is the differing vibration patterns of the incredibly small strings that give rise to quarks, electrons, photons, the force of gravity — in fact, everything in the universe, including us. Numerous string theorists, and especially Ed Witten, as reviewed by Brian Greene in *This Elegant Universe*, have shown that quantum gravity, relativity and supersymmetry are coherent at the most fundamental level and in fact are a natural consequence of the mathematics of superstring theory. A musical metaphor may help. A one-dimensional string on a violin can produce many different notes and overtones, depending upon the length that is bowed or plucked. If a musician blows through the same length of a two-dimensional instrument, such as a flute or organ, the same notes can be played but the overtone vibrations and sound are very different from those of the violin. If the two dimensions are stretched out, as in a xylophone or as a drumhead, instead of being rolled up as in a woodwind or brass instrument, the sounds created are again different. Extending the metaphor to three dimensions, striking metal rods and other percussion instruments creates other, yet more complex sounds. The result is a symphony orchestra, and the music it can play is infinite. String theorists propose that at the moment of creation a single sound (OM!) appeared from nothing, from a quantum fluctuation, and from it separated all the vibrations, notes and harmonies of the universe.

This string-based pathway towards the stars and the theory of everything is eminently satisfying to many physicists, as emphasised by Greene. However, for many others, such as science reporters John Horgan, author of *The End of Science*, and David Lindley, author of *The End of Physics*, there remains a healthy level of scepticism. Physicists working on superstring theory, Lindley contends, are no longer doing physics because their theories can no longer be validated by experiments, but only by subjective criteria, such as elegance, harmony and beauty. Particle physics, he concludes, is in danger of becoming a branch of aesthetics. The physicist Lee Smolin, in *The Life of the Cosmos*, argues that a new view of the world, a new cosmology, is necessary to provide the overall framework for our understanding of Nature. He maintains that "this framework must provide new answers to questions which developments of this past century have made urgent: Why is there life in the universe? Are the laws of physics universal truths, or were they somehow created, with the world? Is it possible to conceive of, and understand, the universe as a whole system, as something more than a sum of its parts?" P.J.E. Peebles, one of the world's most distinguished cosmologists, agrees that it is both an exciting time and a confusing time for cosmology. He says in a *Scientific American* issue (January 2001) featuring "Brave New Cosmos: Can the Universe Get Any Stranger?" the following: "All the ideas under discussion cannot be right; they are not even consistent with one another."

One reason for superstring scepticism is that the theory requires the universe to be described in terms of ten dimensions. This poses a problem. It is hard for most of us to comprehend the addition of seven more dimensions of instruments to a symphony orchestra. Even mathematicians and physicists who think in ten dimensions are bound by bodies that are confined by three dimensions (Figure 1.2) and live a life that begins at a defined time at birth and ends at an uncertain, but all too finite, point in time (Figure 1.3). Thus, the beauty of superstring theory may resolve some of the uncertainty at the boundaries of space and time but it is unlikely to eliminate the uncertainty in our passage through space and time.

A mystery of life is that we can conceive of things that we cannot experience. Conversely, we can experience things, such as beauty and love, which we cannot describe scientifically. For me, and many others, it is the extrapolation of these indescribable experiences that leads to belief in a Creator and an origin of being that lies beyond science.

Accepting Uncertainty

A belief in and knowledge of science requires accepting these uncertainties. My thesis is that it is equally essential that we accept uncertainty in our unique personal relationship with the physical world.

The term I use to express this viewpoint is "quantum theology". Quantum theology is a theology of acceptance. It is especially acceptance of uncertainty. It is acceptance that as a human being composed of matter, there is a finiteness associated with the human condition. It is acceptance that we humans cannot learn or know everything. It is acceptance that we are bound by time: time that we perceive but cannot understand. Our very act of probing God's handiwork or questioning its consequences changes it and makes the results uncertain.

To some people the viewpoint of quantum theology may seem inadequate. They want to believe that it is within our power to understand anything we choose if we have sufficient dedication and perseverance. I believe that such confidence in human beings is unwarranted based both on the history of science and philosophy and on the Revealed Word. The book of Job in the Bible is devoted to our limitations in knowing God. The arrogance of Job's companions some 3,000 years ago in maintaining the certainty of their position was unacceptable to God then. I believe it is equally unacceptable today if a scientist takes an equivalent position that claims certainty.

Einstein's criticism of quantum mechanics is often quoted: "God does not play dice." Quantum theology does not say God is a gambler. It says we must accept that we can only see six sides to a die and that we can only count six numbers. God could have made dice with any number of sides or numbers and may have done so. But we can only perceive the six sides and six numbers of the dice God has chosen to give us. That limitation is ours, not God's. We must accept the limitation rather than lament it. Only by acceptance comes assurance. It may not be the assurance of complete understanding or the certainty we wish for, or even pray for, but it is the assurance that God accepts us even if we have difficulty accepting ourselves and our limitations.

Bibliography

1. *Genesis and the Big Bang*, Gerald L. Schroeder, Bantam Books, New York, 1990
2. *Paradigms Lost: Tackling the Unanswered Mysteries of Modern Science*, John L. Casti, Avon Books, New York, 1989
3. *God and Men*, John Bailey, Kenneth McLeish and David Spearman, Oxford University Press, Oxford, England, 1990
4. *The Edges of Science: Crossing the Boundary from Physics to Metaphysics*, Richard Morris, Prentice-Hall, New York, 1990
5. *The Fifth Essence: The Search for Dark Matter in the Universe*, Lawrence Krauss, Vintage, Basic Books, London, 1989
6. *The Big Bang*, Joseph Silk, W.H. Freeman and Co., 1989
7. *A Brief History of Time*, Stephen W. Hawking, Bantam Books, New York, 1989
8. *Cosmos as Creation*, Ted Peters (ed.), Abingdon Press, Tennessee, 1989
9. *Scientific Creationism*, Henry M. Morris (ed.), Creation-Life Publishers, California, 1974
10. *The Strange Story of the Quantum: An Account for the General Reader of the Growth of Ideas Underlying Our Present Atomic Knowledge*, Banesh Hoffman, Dover Publications, New York, 1959
11. *The Dancing Wu Li Masters: An Overview of the New Physics*, Gary Zukav, Bantam Books, New York, 1980
12. *Modern Cosmology and Philosophy*, John Leslie (ed.), Prometheus Books, New York, 1998
13. *Superstrings: A Theory of Everything?*, P.C.W. Davies and J. Brown, Cambridge University Press, Cambridge, England, 1989

14. *The Limits of Science*, Peter Medewar, Oxford University Press, New York, 1988
15. *Gods and Men: Myths and Legends from the World's Religions*, John Bailey, Kenneth McLeish and David Spearman, Oxford University Press, 1981
16. *About Time*, Paul Davies, Simon and Schuster, New York, 1995
17. *The End of Science*, John Horgan, Broadway Books, New York, 1996
18. *Science and Creationism: A View from the National Academy of Science*, 2nd edition, National Academy Press, Washington, D.C., 1999
19. *The Life of the Cosmos*, Lee Smolin, Werdenfeld and Nicolson, The Orion Publishing Group, London, 1997
20. *The Elegent Universe*, Brain Greene, Vintage Books, Random House, New York, 1999
21. *Modern Cosmology and Philosophy*, John Leslie (ed.), Prometheus Books, Amherst, New York, 1998
22. *Scientific American*, January 2001, pp. 28–47

3
The Origin of Life

We now discuss the current understanding of the origin of life and the evolution of life. The subjects contain strong scientific disagreements and suffer even stronger disagreements from those who believe in a literal interpretation of the Bible. So, again, our goal is to explore the extent of uncertainty in the science of the origins and evolution of life, and compare our knowledge to the creation story in the Bible and various creation myths of other cultures.

Creation Stories

We return to "Genesis" verses 20–31 for the fifth and sixth days of creation.

NRSV Genesis 1: 20

> "And God said, 'Let the waters bring forth swarms of living creatures, and let birds fly above the earth across the dome of the sky.' So God created the great sea monsters and every living creature that moves, of every kind, with which the waters swarm, and every winged bird of every kind. And God saw that it was good. God blessed them, saying, 'Be fruitful and multiply and fill the waters in the seas, and let birds multiply on the earth.' And there was evening and there was morning, the fifth day.
>
> "And God said, 'Let the earth bring forth living creatures of every kind: cattle and creeping things and wild animals of the earth of every kind.' And it was so. God made the wild animals of the earth of every kind

and the cattle of every kind, and everything that creeps upon the ground of every kind. And God saw that it was good.

"Then God said, 'Let us make humankind in our image, according to our likeness; and let them have dominion over the fish of the sea, and over the birds of the air, and over the cattle, and over all the wild animals of the earth, and over every creeping thing that creeps upon the earth.' So God created humankind in his image, in the image of God he created them; male and female he created them. God blessed them, and God said to them, 'Be fruitful and multiply, and fill the earth and subdue it; and have dominion over the fish of the sea and over the birds of the air and over every living thing that moves upon the earth.'

"God said, 'See, I have given you every plant yielding seed that is upon the face of all the earth, and every tree with seed in its fruit; you shall have them for food. And to every beast of the earth, and to every bird of the air, and to everything that creeps on the earth, everything that has the breath of life, I have given every green plant for food.' And it was so. God saw everything that he had made, and indeed it was very good. And there was evening and there was morning, the sixth day."

Genesis 2: 1–3

"Thus, the heavens and the earth were finished, and all their multitude. And on the seventh day God finished the work that he had done, and he rested on the seventh day from all his work that he had done. So, God blessed the seventh day and hallowed it, because on it God rested from all the work that he had done in creation."

Genesis 2: 4–7

> "These are the generations of the heavens and the earth when they were created. In the day that the Lord God made the earth and the heavens, when no plant of the field was yet in the earth and no herb of the field had yet sprung up — for the Lord God had not caused it to rain upon the earth, and there was no one to till the ground; but a stream would rise from the earth and water the whole face of the ground — then the Lord God formed man from the dust of the ground, and breathed into his nostrils the breath of life; and the man became a living being."

Thus, on the fifth day the Bible describes the origin of life as a multitude of living creatures coming forth from the seas, followed by the birds of the air. Early on the sixth day followed cattle, creeping things and wild animals. Finally, at the end of the sixth day, man and woman were created in God's image, with dominion over all living things.

Creation myths from other religions have delightful similarities. Among the people of West Africa, Nyame the sky-god created earth and all the beautiful growing things. Two spirits lived within Nyame — a man spirit and a woman spirit. They landed on the earth one day when Nyame sneezed and blew them out. The woman spirit, whose name translates as Great Mother, eventually got lonely when the man spirit kept going off playing. One day the man-spirit said, "Why are you suddenly so happy?" She answered, "I have a plan to end my loneliness forever. We'll take some clay and mould some small creatures, shaped like ourselves. We'll bake them in the fire and then breathe life into them so they can move about like us. Then they'll be here to keep me company when you go off to the forest. We could call them children."

The man-spirit agreed, so they moulded the first batch of little figures with clay which the woman spirit dug up.

They were surprised when their children were finished. Some were hardly baked at all and were quite white. Others had turned yellow; some were baked red or brown; a few were burned quite black.

The spirit people were delighted with all the children they had made, whatever the colour. They breathed their breath into each of them, so that they came to life just like children waking up from sleep.

The Great Mother was lonely no more. All her children, red and yellow and black and white, continue to worship her even today.

From Nigeria and the Ijaw people comes the creation story of "The Streams of Life":

Mother Woyengi stepped down in a lighting flash from sky to earth, scooped up handfuls of damp earth and shaped it into dolls, lifted each to her nostrils and gave it the breath of life.

She used up all the earth; the dolls covered the plateau. They were brown and naked with blind closed eyes, spindly limbs, no sex, no clothes.

She picked each up, smoothed their eyes, opened and whispered in their ears, asked each one, "You can choose to be a man or a woman for the rest of your life — which will it be?" She dressed them accordingly. They looked around talking. She then spoke, "Your first gift was life, your second gift was sex. For your third gift you can choose the kind of existence you want to live on earth." Some said, "I want a dozen children," "I want to be wise," "Give me a big house," "Let me be fearless," "Help me make music."

Mother Woyengi waited until all were finished and said, "As you've chosen, so it is. Your wishes are granted." And they were.

She then carried them to two streams, rippling across the plain as far as the horizon. Woyengi knelt on the ground between the streams and set her people down. She said, "The stream on this side leads to luxury; the stream on that side leads to ordinariness. You've chosen the kind of life you want; go to the proper stream, and let its water carry you where you chose to be."

The human beings looked at the two streams. Both shimmered placidly in the sunlight. But when the people who had asked for riches, fame or power stepped into their stream, they found it fast-flowing and dangerous with weeds and currents.

The people who had asked for humble, helpful or creative lives stepped into the other stream and found it shallow, clean and clear.

Both sides shouted back their discoveries, and several of the people still on shore began to draw timidly back from the water and ask Woyengi if there was time to change their minds.

Sternly she shook her head. The life they had chosen on the plateau was fixed forever; they had no choice but to go to it.

So, one after another, Woyengi's children began floating or swimming in the stream of riches and the stream of ordinariness, and the waters carried them away and began to irrigate the world with the human race.

From Canada and the Thompson Indians comes the creation myth of "The Five Sisters":

Everything began with water, a glassy smooth sea filling the universe from edge to edge. Above it floated a cloud, and on the cloud lived Old One.

Old One was Chief of the universe. But he was also the only being in existence, and had no company but his own. He got bored.

After a thousand ages he decided to make a world, to create land in the sea and fill it with teeming fascinating life.

First, Old One converted his airy cloud into a heavy fogbank, hovering just above the sea. Then he plucked five hairs from his beard and planted them in the fog like seedlings. He sprinkled them with seawater and they grew into five tall beautiful girls. Old One took them aside one by one to ask what they would like to be. The first said, "I'd like to be Earth, fruitful and fertile, the home of life." Old One said, "So it shall be — all creatures on land will be your domain."

The second said, "I'd like to be Water, rippling everywhere across the earth." Old One replied, "So it shall be — all creatures in the sea shall be yours."

The third exclaimed, "I'd like to be Fire, blazing everywhere, and bring all earth's creatures light and warmth!" "So it shall be," replied Old One. "Your home will be in flints, logs, tree bark, twigs and grass. Your light will be reflected in the sun, moon and stars. Your children will be lizards, salamanders, snakes and insects."

Fire joined Earth and Water on the fogbank's edge. Earth somersaulted from the bank to the sea and at once changed into continents, plains and hills. Her every fold filled with life.

As soon as Earth was formed, Water and Fire dived to join her. Water fell to Earth in a shower of droplets, made pools, rivers and lakes; she stirred it into life.

Fire fell to Earth in a lightning flash and burned herself deep into the hollow caves, in volcanoes with red-hot lava flowing to the surface. Her sparks sealed in twigs and dry leaves waiting to be discovered; in the sky above she left pieces of herself, the sun, moon and stars.

Old One and the two remaining daughters watched these happenings. He asked them, "What would you like to be?"

One answered, "I'd like to be the mother of man and woman. My children will be wise and kind and will live on earth in peace and harmony."

"Rubbish," said her sister. "I shall be mother of the human race. My children will be strong, cunning and pitiless, and will make slaves of all the other creatures in the world."

Old One sighed. "So be it. You've both chosen and your wishes are granted. From now on you'll live on your sisters Earth, Fire and Water, who will nourish your children and give then life."

"There will be both good people and bad people. The world will be full of murder, battle and misery. But I have one present left: the gift of hope. One day, after long centuries of suffering, good will triumph and bad will disappear from earth forever. So I promise and so it will be." So, Old One helped his two daughters to earth and soared into the sky and was never seen again. Creation was finished and humanity's long life began.

There are dozens of creation stories. Nearly every culture has a distinctive version of how life and the diversity of life originated. Most stories provide an explanation for the presence of good and evil, bad people and good people, and eventually the triumph of good over evil. Many characteristics are common to most myths and to the Bible. Life often originates from the sea; clay is often involved in forming people; a serpent or sea monster is many times a central figure in the myth; and the creator breathes the breath of life into the first man and woman.

Requirements for Life

Most cultures provide some mythical explanation for how the breath of life began. Today's scientific culture has life arising from a primordial soup. However, like the big bang theory, the primordial soup theory is not a single unified theory, agreed upon by all scientists. The primordial soup theory has three major versions, which we will examine: the single origin version, the double origin version, and a biomineralogical or clay version. There are major problems with each of these theories.

First, however, we need to ask what the features of life are. What distinguishes a living organism from a non-living object? When does life begin and when does life end? The answers to these questions are central to developing an understanding of the possible origin of life. They are also critical to many of the ethical issues facing today's society. As we discuss in Chapters 6–10, it is not possible to understand the effect of technology on the preservation of life without answering these questions.

Most biologists agree that there are three primary requirements for life to begin and be maintained. These requirements are self-replication, self-maintenance and self-adaptation. All living organisms known today, all 2,000,000–3,000,000 species, from the single-celled amoeba to humans containing billions of cells, have these three features. There are no exceptions.

Self-replication is the ability to make a copy of oneself. All living things have this ability. Cells divide and the two new cells are generally indistinguishable from the original cell. Ageing and disease occurs, in part, when this copying process starts slowing down and making mistakes. All cells known today replicate by the same mechanism. The information needed to reproduce a cell is in the form of DNA, contained within the cell. The information required to organise cells into structures or tissues in multi-celled life forms is also stored in DNA.

One of the great triumphs in biology was Watson and Crick's discovery of how information is stored in the double helix DNA molecule. It is by the nucleic acid base pairs A–T and G–C, adenine paired with thymidine and guanidine paired with cytosine (Figure 3.1). These four nucleic acids, attached structurally to deoxyribose, a sugar, and a phosphate group, are the letters of the biological alphabet which stores the information. The information is passed on, in a process called transcription, from generation to generation of cells by the strands of DNA being replicated in the form of RNA. RNA is slightly different structurally from DNA, with ribose sugar (R) instead of deoxyribose (D) in the backbone of the helix; but the information content in the related base pairs is the same. This method of replication of information is very efficient and almost free of errors. All cells of all living creatures known today multiply by division using this method of replication and can do so for many, many generations without significant errors.

Replication of life by means of RNA and DNA began a very long time ago. Radioactive dating of rocks indicates that the earth is about 4.5 billion years old. Scientific evidence favours the theory that the earth was formed by condensation of the heavy elements created in the explosion of stars. It took several billion years from the initial creation events of the universe for multitudes of stars to form, collapse and explode releasing large quantities of heavy elements. The earth and all rocks and life in the earth are composed of the elements made in stars. Thus, each of us is literally composed of

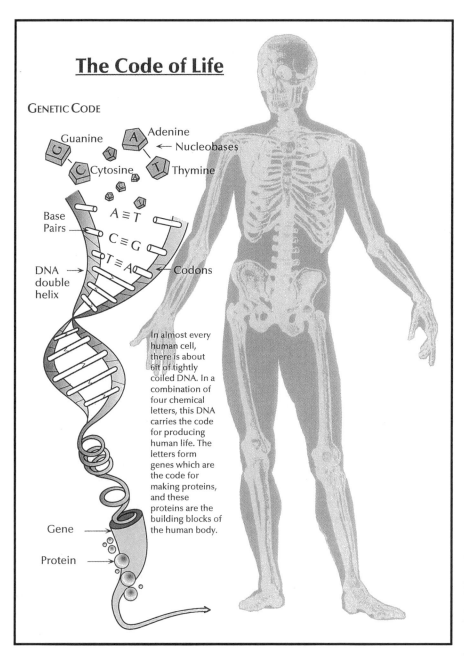

The Code of Life

GENETIC CODE

Guanine

Adenine

← Nucleobases

Cytosine

Thymine

$A \equiv T$

Base Pairs

$C \equiv G$

DNA double helix

$T \equiv A$

← Codons

In almost every human cell, there is about 6ft of tightly coiled DNA. In a combination of four chemical letters, this DNA carries the code for producing human life. The letters form genes which are the code for making proteins, and these proteins are the building blocks of the human body.

Gene

Protein

Figure 3.1 The genetic code of life.

"star dust". When we die we return as dust again. This is not myth: it is reality.

Paleontological evidence from fossils found in rocks indicates that primitive single-celled organisms came into existence approximately three billion years ago. Although data are sparse, these early life forms appear to have had the same DNA–RNA method of information storage and replication as the life forms known today. Thus, to paraphrase "Genesis", "Self-replication was created and it was good." Other methods may have been tried, such as that suggested by the Cairns–Smith intelligent clay theory, which I will come to later, but were discarded as inefficient billions of years ago. Answering the question as to how self-replication could have arisen from the primordial soup is central to a scientific theory of the origin of life.

The second requirement for life is self-maintenance. For an organism to live it must be capable of ingesting food and converting the energy content of the food into cellular processes like movement, replication or elimination of by-products. The conversion of foodstuffs into useable energy to maintain life is called metabolism. The energy from metabolic processes is also used to construct proteins, the building blocks of the cells, and for cell growth and cell replication. All living matter has metabolic processes linked together with replication processes within the cell wall. The two requirements are inseparable. Life requires both.

The third requirement, self-adaptation, is necessary for a life form to be maintained for many generations.

Adaptability is an essential requirement for life because the external environment is constantly changing. If an organism is not capable of adjusting to environmental changes, it will die. If none of its relatives can adapt they too will die and the species will become extinct. Many species have died out in our lifetimes and many more will do so within the next few decades. Untold numbers of species have died out since life first appeared on earth. The life forms that survived did so because they could change as the earth's atmosphere and biosphere changed. They survived because they could adapt to the environment.

Adaptability does not come from a single cell or a single organism. It comes from within a population of organisms. When a cell divides there is a possibility that the replication process will produce a new cell that is slightly different from its parent. If this difference is detrimental the new cell may die and the difference will not be passed on to later generations. However, if the environment has changed, the difference in the new cell may allow it to survive when others without it do not. The difference, or adaptation, is then passed on to later generations via self-replication. Eventually all the organisms will contain this difference and the old ones will be gone. That is partly why there are no dinosaurs, or mastodons, or sabre-toothed tigers, or even platypuses or dodos around today, except in movies.

The DNA–RNA method of self-replication results in self-adaptation because small changes in base pairs along the DNA helix can produce differences in cell behaviour and cellular response to the environment. The objective of the multi-billion-dollar human genome project, which we will discuss later, is to determine the sequence of the 2–3 billion base pairs in the DNA strands of humans. This understanding should show why genetic defects occur in human beings and how to prevent them. In Nature, negative genetic defects usually die out on their own because the organism is at a disadvantage and cannot compete and survive. In today's world, we can keep beings with life-threatening genetic defects alive. It is an ethical decision whether we should do so for the good of the human race. We will discuss these issues in later chapters.

In this chapter, the problem we address is how self-adaptation could have been created within a life form on primitive earth. It is a complex, triple problem. The triple problem is that all life forms known to exist or to have existed on earth contained self-replication, self-maintenance and self-adaptation simultaneously within their cells. How could this three-fold requirement for life have been achieved from a non-living environment? There is no easy answer. There is hardly any scientific answer at all. There is only a primordial soup theory.

The Primordial Soup Theory

The primordial soup theory begins with an assumption that is similar to many creation stories: that life formed first in the seas. However, the seas were very different three or four billion years ago from what they are now. The earth's surface was very hot, with deep cracks created during shrinkage of the molten ball as it cooled to form a crust on the earth. Volcanoes erupted frequently, spewing forth methane, ammonia, water vapour, sulphur and hydrogen gases produced deep within the earth. The atmosphere was highly reducing and low in oxygen. Temperatures fluctuated widely and water vapour condensed during the night, forming pools. The water dissolved elements such as sodium, potassium, calcium, magnesium and even iron and aluminium from the hot rocks. Hydroxyl, carbonate and phosphate anions were also concentrated in the alkaline ponds. The gases in the atmosphere were continuously exposed to high voltage electrical discharges from lightning, to high energy particles from intense radioactivity, and continuous bombardment by high energy ultraviolet radiation from the sun since there was no protective ozone layer at that time.

Exposure of methane, ammonia, water vapour and hydrogen to the high energy particles formed bonds between the very small molecules, thereby creating new, larger and more complex molecules. Some of the types of molecules found in living organisms were created by these chemical processes.

When the temperature cooled at night, water condensed and formed a fog or rain. The new molecules condensed in droplets with the water and fell to earth, gradually forming more and more highly concentrated solutions in ponds, grottoes or cracks. Thus, the primordial soup was formed (Figure 3.2).

The first problem with the primordial soup theory is that it assumes that all the building blocks of life were condensed into these puddles or soup bowls. A long list of molecules must be present in the soup to form life. There must be simple amino acids, at least eight different types, according to Dyson, to form even the simplest

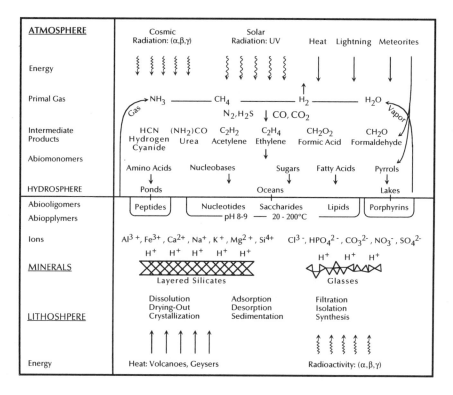

Figure 3.2 Schematic representation of the primitive earth as a chemical laboratory for the synthesis of biological molecules on the surface of bioactive minerals and glasses. (Modified from W. Schwemmler, *Reconstruction of Cell Evolution*, CRC Press, Boca Raton, Fl, 1984.)

proteins found in cells. There must also be carboxylic acids, nucleobases and sugars to form the base pairs for a primitive type of DNA. In addition, there must be fatty acids to form membranes. It is important to remember that the three requirements for life discussed above could not be achieved unless all the above constituents of life were present together, or occurred in a highly specific orderly sequence.

Although the concept of a primordial soup had been suggested for many years by a number of people, its status as a standard theory of science was achieved by the results of an experiment by a graduate student, Stanley Miller, at the University of Chicago in 1953. Miller, working with a reluctant OK from his Nobel laureate mentor, Professor Urey, performed what is often cited as the "critical experiment". He passed a mixture of gases simulating the reducing atmosphere of primitive earth through an electric discharge that simulated lightning bolts.

After several tries he found in the flask what he had been looking for — amino acids. There were only a few of the simpler amino acids present, but one of them, alanine, was in surprising abundance, of almost 3%. Miller, and others using similar methods, in the years since also found in the flask other compounds that are often called the "building blocks of life". The primordial soup theory assumes that these molecules, formed in the atmosphere by electric discharges, then rained upon the earth, and became concentrated in ponds where they could interact chemically to form the large molecules and structures of primitive cells.

However, there is a big problem with the theory. No one has ever produced a living organism of even the simplest type from this type of test tube experiment. In fact, no one has even come close. Nearly 50 years of effort have yielded not even a segment of a DNA molecule from simple molecular precursors. The three requirements for life have never been achieved experimentally without first starting with constituents already created by life forms.

The old cliché of which came first — the chicken or the egg? — is symbolic of the problem with the primordial soup theory. The scientific version of the chicken-versus-egg question that is relevant to the origin of life is as follows: which came first — proteins (required for cell metabolism) or DNA (required for cell replication)?

Three alternatives to the primordial soup theory have arisen to get around this problem. The single origin version of the theory has replication and metabolism linked irreversibly together within a cellular membrane from the beginning. A single origin advocate, such as Oparin, argues that primitive cells must have originated first with characteristics similar to the drop-like materials he has studied, called coacervates. In such versions, the primitive cells did not have a complicated replication apparatus, like DNA. Instead they grew and divided largely by chance. Later on, it is hypothesised, simple proteins composed of amino acids formed enzyme-like structures which evolved within the membranes. The proto-enzymes organised random populations of molecules into repeating metabolic cycles. Eventually, after millions of years of trial and error, strands of RNA appeared by chance. The RNA self-organised into primitive genes, forming DNA, and the now omnipresent nucleotide-based replication slowly evolved to dominate forevermore. You might say this is a "chicken first, egg later" version.

In contrast, Eigen's single origin version has genes first inside the membrane with enzymes forming second, and cells appearing third. In other words, an "eggs first, then chickens" version.

A fundamental problem with any of the single origin theories of the origin of life is the highly improbable situation of metabolic and replicative processes appearing at the same time in a membrane, just by chance. Shapiro discusses the probability of these complexes occurring only by chance. The calculations suggest that for this to happen it's about as probable as letting a pot of chicken soup cool overnight and the next morning finding a chicken egg or a baby chick chirping away in the pot.

In order to improve the probability, numerous scientists suggest that life originated in a two-step, or double origin, process. In this version, two types of pre-living organisms formed in the primordial soup — one type capable of metabolism without replication and the other capable of replication without metabolism. For many millions of years, these primitive molecular organisations are proposed to have formed, grown, dissolved, and re-formed from the soup, neither of them living and satisfying the three requirements for life.

However, at some point in time about three billion years ago, it is assumed that the two structures fused together inside a common membrane and life began. This process is called symbiosis. Symbiosis is when two organisms' survival depends on each other. For example, the bacteria in your gut that are digesting your most recent meal are symbiotic. Their survival depends on your eating. Conversely, you will not get the best from your food without their help. So, they benefit from you and you benefit from them. Symbiosis is believed to be the way the mitochondria of oxygen-processing cells, such as those that make up us, became incorporated within anaerobic cells, like bacteria, about two billion years ago.

The double origin version gets around, to some extent, the problem of synthesising a DNA or RNA double helix from simple molecules, which as yet has not been possible in the laboratory. In the double origin approach, only amino acids have to be formed in the primitive cell, which is a much simpler process. Thus, in the double origin theory the nucleotides that eventually became DNA and RNA formed as a by-product of protein metabolism. This version most closely matches the calculational requirements for life to form from a co-operative organisation of a large population of molecules, as described by Freeman Dyson.

However, a major disadvantage in double origin versions of the creation of life is the uncertainty as to how a parasitic RNA-based replicating, but not metabolising, organism could ever have arisen and survived long enough to establish the symbiotic relationship with a non-replicating protein-based life form. There is no scientific evidence to indicate that this step is feasible, let alone demonstrable.

The third version, proposed by Cairns-Smith, is also a double origin theory. However, instead of the first step being protein-based he proposes it was more likely to be clay-based. Clay is composed of very small crystals made up of ordered sheets of aluminium and silicon ions, bonded together by oxygen ions. The clay crystals are formed by the breakdown of the minerals that make up the Earth's crust. The highly regular pattern of ions holding together the crystals is suggested to be the key to creation of biological molecules; it is the key to life in this version.

The clay crystals are proposed to have served a replicating function in the primitive cells similar to the role of DNA in today's cells. In other words, protein synthesis — which is the assembly of disordered random organic molecules from the primordial soup into a highly specific ordered structure — occurred on a template of clay crystals. Electrical charges on the clay crystals acted somewhat like the bumps on Lego blocks. They fixed the order of the organic building blocks that attached to them so that patterns repeated accurately over and over again.

This bio-mineralogical version is the most spiritually satisfying because of the centrality of clay as the substance from which man is formed. As we saw earlier, this is common to many creation myths. It also coincides with the evidence that the earth existed in an inorganic, mineral form for about two billion years before life appeared. So a transition of ordered living life forms from ordered non-living substrates seems reasonable, at least superficially.

The huge difficulty of the clay-based version of the primordial soup theory is best repeated by Dyson, who says, "There is no experimental evidence to support the statement that clay can act either as a catalyst or a replicator with enough specificity to serve as a basis for life." In other words, if clay can provide sufficient ordering for organic molecules to create RNA, it has not been demonstrated in any laboratory to date.

Science requires demonstration of evidence and proof by repetition for a theory to be accepted. Our belief in science depends upon this fact. It is what distinguishes science from myth. During the last 3,000 years countless scientific theories have been abandoned in the light of new experimental evidence. The scrutiny and testing of the primordial soup theory does not measure up to what is required to call it a truly scientific theory. None of the three versions outlined above have been demonstrated in the laboratory. There is also no conclusive evidence for either of the three versions in the sedimentary deposits. The most we can say about the science of the origin of life is that several unproven but scientifically plausible hypotheses exist.

Some years ago I published another, also unproven, version of the primordial soup theory. This version, termed the bioactive substrate theory of the origin of life, is based upon experimental evidence obtained in our studies of the molecular mechanisms of the bonding of bioactive glasses to living tissues. In its favour is some experimental basis for the theory. My interest in origins arises from the philosophical implications of the experimental — and now clinical — fact that certain man-made compositions of bioactive glasses bond to bone. The details are discussed in Chapter 5. The intriguing question posed by this discovery is: "How can a man-made substance bond to a living substance?" Or, in a philosophical sense, "How can a man-made and a God-made substance become one?"

In trying to answer this question the first important point to recognise is that "humans uncover rather than discover". By this I suggest that the principles underlying our scientific discoveries have been in existence since time began, 7–20 billion years ago. We humans only uncover these principles as we make ever broader and more thorough observations. There is in fact a limit to how much we can uncover, as we saw in Chapter 1.

The implication of this observation is that the physical-chemical and biochemical factors responsible for the bonding of Bioglass® to bone may already have been in existence before I made the first glass compositions in November 1969. When the composition was right the glass bonded. When the composition was wrong the glass did not bond. Somehow the tissues of the rats, the experimental animals used in the first tests of Bioglass®, must have been pre-disposed to form a bond to the glasses by aeons of evolution. One cannot trick Mother Nature. The DNA code must include the information which permits these specific compositions of glasses to create a bond to living tissue. In Chapter 5 we discuss the genes that are activated by these bioactive glasses and their dissolution products.

The conclusion is: "The principles that govern the bonding of living to non-living matter are related to those responsible for the creation of living matter from non-living matter." This conclusion is a big step to take but a logical one. It is a step that can lead from disorder to order. It is a step that can lead to life.

Disorder ⇌ order reactions are one of the most important classes of phase transformations, i.e. gas ⇌ liquid, liquid ⇌ solid, glass ⇌

crystal. The free energy of a system is lowered in each case at the expense of entropy, or vice versa, depending upon the direction of the transformation. Biological systems fit into the category of order ⇌ disorder reactions. A biological phase transformation has profound consequences; with order comes life, with disorder, death.

Life transforms disorder into order. Living organisms transform a random mixture of organic and inorganic molecules into highly organised macromolecular structures. Death is the onset of the reverse reaction:

$$\text{Disorder} + \text{Life} \rightarrow \text{Order} \;\; (\text{Living}) \tag{1}$$

$$\text{Order} + \text{Death} \rightarrow \text{Disorder} \;\; (\text{Non-living}) \tag{2}$$

As required by the third law of thermodynamics, life is achieved and maintained at the expense of the entropy of its environment. The mystery of life is the origin of this transformation; the mystery of death is what stops it.

For non-living systems the physics of a phase transformation is straightforward; changes in pressure, temperature or volume modify the interaction of electron wave functions of neighbouring atoms, which leads to a lower, or higher, total energy for the assembly of atoms. A significant feature of this physical process is its reversibility. As the term implies, invariant points on *Temperature (T) — Pressure (P) — Composition (C)* diagrams are independent of the direction of heating or cooling. Although kinetic barriers to order ⇌ disorder transformations exist, thereby permitting, for example, glass formation, equilibrium requires that all supercooled or superheated systems eventually and reversibly transform. A crystal can be cycled through its melting point T_m innumerable times and the same space group will always reform upon crystallisation.

The phase transformation of a living system is remarkably different. The death of a cell is irreversible. Chemical analysis will show that the same molecules are present in the dead cell; even the supra-molecular architecture of the cell wall, nucleus and organelles is still present. But entropy (disorder) irreversibly

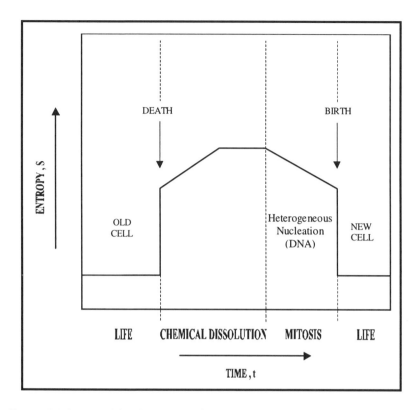

Figure 3.3 Irreversible effect of death on disorder (entropy) of a living cell.

increases following death (Figure 3.3). There is no change in the temperature, pressure, or chemical environment that will resurrect vitality in a dead cell. The biological transformation from order to disorder is one way:

$$\text{Life (Order)} \rightarrow \text{Death (Disorder)} \tag{3}$$

Restoration of life requires decomposition of the organism and regrowth, by mitosis (cell division) of new living cells (Figure 3.3). This disorder $\rightarrow$ order transformation to form a new cell is not self-nucleating. The concept of spontaneous or "homogeneous" nucleation of life is called a "miracle" and is in the domain of myth rather than physics. All aspects of the formation of a new cell are directed by the base pair sequences in DNA.

Thus, it is accepted that the disorder → order phase transformation characteristic of life involves the equivalent of "heterogeneous" nucleation by DNA.

It is equally well accepted that all DNA is synthesised within a cell, directed by the DNA already present. No DNA is synthesised *de novo* outside a cell and transported into a cell to service its replicative function. Even viruses, which can be considered as mostly "naked" segments of DNA, are replicated within a host cell using the biosynthesis pathway already present in the living cell.

The Paradox

The problem of understanding the origin of the disorder → order transformation of biological systems takes the form of a paradox. It is two versions of the same paradox we encounter in establishing a theory for the origin of life:

(1) DNA synthesis requires DNA. How did DNA biosynthesis begin without DNA?
(2) Likewise, protein synthesis requires enzymes. Enzymes are proteins. Thus how did protein biosynthesis begin without enzymes?

In other words, "How could the transformation reaction in Eq. (1) (disorder → order) occur?"

None of the three versions of the primordial soup theory or their modifications provides a mechanistic solution to surmounting the entropic barrier (illustrated in Figure 3.3) that is necessary to form an ordered assemblage from a disordered one. The "life from space" conjecture of Hoyle and Wickramsinghe simply shifts the question to an extraterrestrial origin without providing a means of proof. It is not science, although recent interstellar spectroscopy confirms the conjecture that biomolecular synthesis of simple molecules may be possible in space. The other three hypotheses summarised above also offer no means of proof. The macromolecular structures characteristic of life cannot be produced by non-directed synthesis.

The bioactive substrate hypothesis for the origin of ordered biotic structures from a random assemblage of disordered organic molecules provides a means of achieving ordered arrays from a disordered assembly of biological molecules.

This hypothesis suggests that selective chemical adsorption of biomonomers may have contributed five ordering factors that eliminated randomness, reduce entropy, and ensure repeatability of the polymers formed. The ordering factors are:

(a) Substrate steric factors which impose repeatable spatial requirements;
(b) Monomeric optic axis orientation which imposes chiral growth;
(c) Substrate optic axis orientation which imposes a match of monomeric chirality;
(d) Irreversible condensation reactions which result in a stable biopolymer;
(e) Polymeric steric factors which limit the selection of additional monomeric units.

Evidence for Bioactive Substrates

Our original evidence in support of a bioactive substrate theory for the disorder → order transformation was the epixatial binding of the levorotatory (left-handed) form of the amino acid alanine (poly-L-alanine) to alpha quartz crystals. Highly specific orientations of poly-L-alanine were formed on $(10\bar{1}0)$ or $(10\bar{1}1)$ crystal planes where repeating silanol groups matched alanine binding sites [Figure 3.4(a)]. The expitaxially oriented amino acids were tightly bonded and resisted mechanical abrasion or chemical attack, unlike poly-D-alanine or substrates where spatial distances between bonding sites were not matched. These experiments provided evidence for ordering factors (a), (b) and (c) above and perhaps (d) because the adherent agglomerates could not be removed from the substrate without substantial mechanical force.

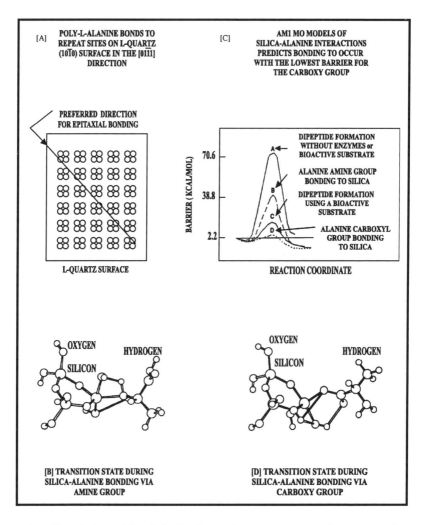

Figure 3.4 Chemical models of the binding of amino acids to a bioactive substrate.

An Inorganic Route to Biosynthesis

Based upon these empirical findings, a research colleague, Jon West, and I completed several quantum calculational studies. The calculations indicate a possible mechanistic solution to the order ⇌ disorder paradox of life. Our first calculations studied the energetics of interaction of hydroxylated SiO_2 clusters with water and with amino acids. Our rationale is that silica, silicates and water have always been the most abundant compounds on the surface of the Earth and provide the most probable interfaces for interactions with pre-biotic (non-living) amino acids.

The calculations show that hydrated three-membered silica rings are easily formed during the fracture of silica and silicate rocks of the Earth's surface. The three-membered silica rings are energetically metastable due to quantum-mechanical strain of the bridging Si–O–Si bonds. The strained three-membered silica rings provide a pathway for the binding of alanine amino acid molecules [$CH_3CH(NH_2)COOH$] with an energy barrier as low as +2.2 kcal mol^{-1} [Figures 3.4(b) and (d)]. The low energy barrier is easily reversible at 25–50°C and is relatively insensitive to hydrolysis conditions of the molecules. The low energy barrier is due to the formation of a pentacoordinate Si atom in a metastable transition state. The pentacoordinate silicon state occurs when the –COOH group of an amino acid interacts with a trisiloxane ring. The carboxyl bond polarises the Si–O–Si bond in the three-membered ring and opens it into a three-membered chain. Water attacks three-membered rings in a similar manner, as shown experimentally and theoretically using hydrated silica gel systems. (This equivalence of the polar behaviour of carboxyl groups of amino acids to H_2O is vital when we examine the bonding of bioactive glasses to living tissues in Chapter 5.)

Our quantum mechanics calculations also show that the metastable pentacoordinate Si–OH complex acts like an inorganic enzyme by providing a favourable reaction pathway for polypeptide (protein) synthesis. The easily reversible opening and closing of the hydrated silica rings provides a pentacoordinate Si transition stage which serves an enzymatic function by providing a low energy pathway to create the dipeptide bond. The calculations show that the addition of a second amino acid (glycine) to a trisiloxane + alanine cluster results in the formation of an alanine–glycine dipeptide and the release of a trisiloxane chain. The energy barrier of the saddle point for the inorganic catalysed reaction of alanine + glycine + trisiloxane is +17.9 kcal mol^{-1}. This barrier for polypeptide formation via inorganic biosynthesis is greater than enzyme-catalysed peptide synthesis. However, this value is substantially less than the barrier to the formation of peptide bonds without a common intermediate. Figure 3.4(c) shows that the energy barrier for formation of a dipeptide bond without the presence of an enzyme-like structure is very high, +70.6 kcal mol^{-1} calculated with the same method.

These calculations demonstrate a possible energetically feasible pathway for achieving the fourth ordering factor (d), i.e. the irreversible condensation reactions involved in protein synthesis, prior to the existence of enzymes.

The calculations also show that the metastable silica–amino-acid transition states have an intense absorption band in the 220 nm region of the ultraviolet part of the electromagnetic spectrum. This is a significant finding, because one of the most likely explanations for the chirality of life forms involves circularly polarised sunlight in this region of the spectrum, as reviewed by Mason in *Chemical Evolution*. Living organisms contain only levorotatory (L) amino acids in proteins. When a racemic mixture of amino acids is exposed to right circularly polarised UV photons in this range of energy, the preponderance of amino acids that survive is levorotatory.

The L amino acids exposed to the metastable surface transition states of silica and silicates would be preferentially chemisorbed on the substrates due to ordering factor (b). The UV absorption at 220 nm would provide photocatalysis reactions of the peptides on the bioactive substrates, thereby stabilising the L amino acids in the proteins. The chirality of the photocatalysed reaction would be maintained by selective polymerisation of only the monomers that satisfied polymeric steric factors, ordering factor (e).

Thus, before the existence of enzymes or DNA the entropic barrier for disorder → order (non-living → living) illustrated in Figure 3.3 would be overcome by the five ordering factors catalysed by the bioactive substrate, as indicated in Figures 3.4 and 3.5.

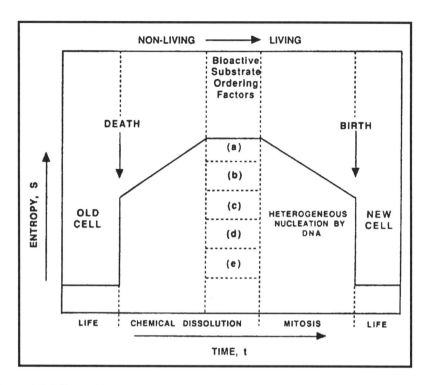

Figure 3.5 Effects of bioactive ordering factors (a–e) overcoming the entropic barrier in creating life.

Lowering the entropic energy barrier (Figure 3.5) combined with a low enthalpy of reaction of amino acids with metastable surface states (Figure 3.4), and photocatalysis, leads to a plausible reaction mechanism for protein synthesis and the disorder → order reaction required to initiate self-propagating life forms from a non-living mixture of amino acids.

The combination of specificity and variety of bioactive substrates may have provided a basis for what have become highly specific and enormously variable organic structures capable of being replicated over and over again. We can conclude that the inorganic origin of biopolymers may be irreversibly and immutably locked into the very beginning of the genetic code.

These inorganic ordering factors provide the irreversibility of bonding between organic molecules and impose a 3-D ordering on the resulting structures that could form from the primordial soup. An important consequence of this theory is the creation of complex organic structures composed of left-handed amino acids and right-handed sugars, a characteristic common to all life forms known today or in the past.

Present Day Implications

Although my version of the primordial soup theory has some theoretical basis and some experimental evidence in its support, it is far from being a proven scientific theory. The same can be said for any other version of the primordial soup theory.

James Gleick says in his best-selling book *Chaos*, "Pattern born amid formlessness: that is biology's basic beauty and its basic mystery. Life sucks order from a sea of disorder." Many years ago Schrödinger, a founder of quantum theory, said, "A living organism has the astonishing gift of concentrating a stream of order on itself and thus escaping the decay into atomic chaos."

A bioactive substrate theory is intriguing because the infinite variability of order and disorder at the interface of the building blocks of the Earth's surface — sand, silica (SiO_2) and water (H_2O) — could well have served as a strange attractor to bring order from chaos — and thus be the materials that have shaped our past.

However, the emerging science of chaos shows that the presence of a strange attractor signals the existence of pure disorder. No point or pattern of points ever recurs. Yet it also signals a new kind of order.

This is an order that is infinitely changeable but at the same time is an infinitely repeatable pattern of behaviour. It is an order that always looks similar but is always just a little bit different. This is also the pattern of life; every individual of a species is the same but every individual is also a little bit different. Uncertainty is always mixed with certainty. Even the mixture is uncertain. To paraphrase Gleick, "Uncertainty is both the beauty and the mystery of life."

Thus, once again, as is the case with the big bang theory, we are left with uncertainty. This does not mean that we should ignore science in these areas or dismiss an interest in determining answers to questions that are infinite in Nature. It is in our nature to seek answers to such questions. It is in such seeking that the infinite acquires meaning. However, it is equally important to recognise and accept uncertainty in such subjects. Our life is filled with uncertainty. Science does not necessarily eliminate that uncertainty at the deepest levels of understanding of life, or death. The significance of this uncertainty becomes clear in Chapters 9 and 10 when we discuss the difficulty of defining human life and death. Our limitations in understanding the scientific boundaries between living and non-living become even more important when we debate the ethical issues associated with genetic manipulation of life and intervention in the beginning and end of human life. Science has made such interventions possible but it cannot predict the consequences. Until we know how life was created we should be ever so cautious about the extent to which we manipulate it. We need to remember that science cannot replace faith — or hope.

Bibliography

1. *Origins: A Sceptic's Guide to the Creation of Life on Earth*, Robert Shapiro, Penguin Books, 1988
2. *Origins of Life*, Freeman Dyson, Cambridge University Press, 1985
3. *Wonderful Life*, Stephen Jay Gould, W.W. Norton and Co., 1989
4. *Paradigms Lost: Tackling the Unanswered Mysteries of Modern Science*, John L. Casti, Avon Books, New York, 1989
5. *Chaos: Making a New Science*, James Gleick, Viking Press, New York, 1987
6. *Bioceramics and the Origin of Life*, Larry L. Hench, *Journal of Biomedical Materials Research*, Vol. 23, pp. 285–303, 1989
7. *Hen's Teeth and Horse's Toes*, Stephen Jay Gould, W.W. Norton and Co., New York, London, 1984

8. *Gods and Men: Myths and Legends from the World's Religions*, J.R. Bailey, K. McLeish and D. Spearman, Oxford University Press, Oxford, England, 1990

9. *Evolution From Space*, Fred Hoyle and N.C. Wickramsinghe, J.M. Dent and Sons, 1981

10. *Reconstruction of Cell Evolution: A Periodic System*, Werner Schwemmler, CRC Press, Florida, 1984

11. *The Wonderful Mistake: Notes of a Biology Watcher*, Lewis Thomas, Oxford University Press, 1988

12. *The Ascent of Man*, J. Bronowski, Little, Brown and Co., Boston/Toronto, 1973

13. *Chemical Evolution*, Stephen F. Mason, Clarendon Press, Oxford, England, 1992

14. *Life and Death: The Ultimate Phase Transformation*, Larry L. Hench, *Thermochimica Acta*, 280/281, pp. 1–13, 1996

4
The Evolution of Life

We have examined the boundary between science fiction and science fact, between certainty and uncertainty, for two of the great questions — the origin of matter and the origin of life. Our next goal is to do the same for a third, equally great question — the origin of diversity of life.

The difficulty with this question is twofold. We need to understand the origin of multiplicity as well as the origin of the complexity of life forms. Millions of different species of living organisms exist today and untold millions of species have lived, died and become extinct since life on the Earth began some 3.5 billion years ago. How did this enormous diversity begin? How has this diversification of life multiplied over the millennia?

Along with the expanded numbers of differing species has come an ever-larger expansion of the complexity of life forms. The difference in organisational complexity between single-celled organisms, such as bacteria and human beings, is staggering. Figure 4.1 illustrates the magnitude of the difference by comparing the approximate information contained within inanimate states of matter and various life forms. The atomic number of the elements of matter (their number of protons) and their atomic weight (the number of protons and neutrons) of each of the 98 naturally occurring elements was fixed at the time of their creation in certain types of stars, as summarised in Mason's comprehensive volume *Chemical Evolution*. Most of these numbers are fixed for all time; the exceptions are the radioactive elements that undergo decay and transformation to elements of lower atomic weight at predictable rates.

The information contained within simple inorganic molecules composed of elements chemically bonded together, such as SiO_2 or

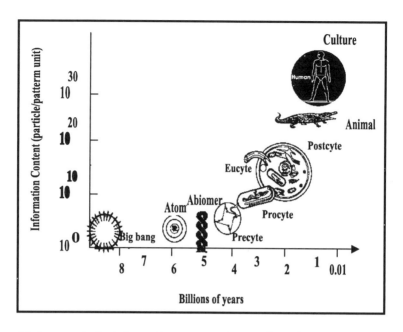

Figure 4.1 Increase in the information content of matter from the origin of the universe to the present.

Al_2O_3, increases as the number of elements in the molecule increases. This complexity is also frozen in time. However, as the complexity of inorganic crystals increases they become increasingly vulnerable to change, by weathering and decomposition. Thus, for example, alumino-silicate-based rocks in the Earth's crust gradually break down by the action of water into hydrated clays. The inorganic chemical reactions that occur in nature lead to states of matter that generally have less order and complexity. These irreversible processes have continued throughout six billion years as the Earth has cooled, and have led to the erosion of mountains, carving of rivers and formation of beaches.

In great contrast, Figure 4.1 illustrates that the effect of time on life has led to an ever-increasing complexity of living organisms. All life forms today have an enormously expanded information content over the inanimate world. Even a simple one-celled bacterium, such as *E. coli*, which lives by the billions in our large intestine, contains 4.6 million nucleotide base pairs in its genetic material. A single cell of a human being contains about 6 billion (6×10^9) base pairs in the DNA of our genes. It is estimated that a human is made up of more than 2 billion cells (2×10^{12}). Thus each person contains within his or her body a collective information content of $6 \times 10^9 \times 2 \times 10^{12}$ or 10^{22} base pairs. This huge figure has a startling similarity to the scale of the limits of distance and time discussed in Figures 1.2 and 1.3.

This enormous amount of stored information is passed on to our children. Thus, as the entropy (disorder) of the inanimate world increases, the living world propagates order and maintains equilibrium.

How did a living being with a cosmic magnitude of order and information content come into existence? The 20th to 31st verses of "Genesis" provide an answer for millions of people worldwide.

> "And God said, 'Let the waters bring forth swarms of living creatures...' There was evening and there was morning, the fifth day.
>
> And God said, 'Let the Earth bring forth living creatures of every kind...' And God saw that it was good. Then God said, 'Let us make human kind in our image, according to our likeness...' And it was so, and indeed it was very good. And there was evening and there was morning, the sixth day."

Many believers in the Bible as the Revealed Word of God accept these verses as the literal truth. They accept that all living species observed by humans today, in the seas and on the land and in the air, were created as unique species by God the Creator. They also believe that all the extinct species found in the fossil record were placed there as part of the Creation, which occurred during six days some 6,400 years ago. This belief in the literal truth of the events described in "Genesis" has become known by its proponents as "scientific creationism".

Scientific creationists are vigorously opposed to the concept of evolution as a means of explaining the multiplicity and diversity of life. During the summer of 1999 the debate between creationists and evolutionists reached disturbing proportions in the United States, when the School Board of the State of Kansas ruled that it was no longer required that a student show an understanding of evolution to obtain a high school diploma. The ruling left schools in the state free to determine whether they would or would not include the teaching of evolution as part of the biology curriculum. The decision led to an outpouring of rage from scientists, columnists and editorial writers from coast to coast. Pro-evolution headlines such as "Creation Views Outrage Top Scientists", in *USA Today*; "In the Beginning, It's a Fact: Faith and Theory Collide over Evolution" and "Wilful Ignorance on Evolution", in the *New York Times*, were countered by large advertisements in major newspapers by pro-creationist groups. For example, an advert in *USA Today*, entitled "The Truth About Evolution", included the following:

> "If this book (*Refuting Evolution*, by J. Safarti) were to get into the hands of students and parents across the nation, there would be a public outcry against evolutionary brainwashing. This indoctrination has an underlying connection to the collapse of morality and increasing school violence across America."
>
> — Ken Hom, *USA Today*, August 27, 1999

On the surface this may appear to be a harmless debate between science and religion, continuing a tradition that stretches back in time to Galileo. However, the content in the above-quoted advertisement lays responsibility for events such as the massacre at Columbine High School in Colorado at the door of Charles Darwin. This is an irresponsible and dangerous assertion. Nearly all scientists will agree that the deranged act of two high school students had nothing to do with being exposed to the principles of natural selection presented by Charles Darwin nearly 140 years ago.

Before we discuss the ethical issues associated with the debate, assuming there are some, it is useful to review briefly the level of certainty in the theory of evolution as a scientific explanation for the multiplicity and diversity of life.

In 1859 the evolution revolution in biology began with the publication of *On the Origin of Species by Means of Natural Selection* by Charles Darwin. He described for the first time a simple systematic mechanism by which both multiplicity and diversity of life forms could, and did, develop. He called the scientific principle "the theory of natural selection". The theory in its simplest form is summarised as "survival of the fittest".

In the previous chapter we saw that there are three requirements for life: self-maintenance, self-replication and mutability (adaptability). Darwin's theory of natural selection rests on the importance of this third criterion for life, i.e. mutability. All populations of life forms from microbes to man compete for food, the basis of metabolism, and for the opportunity to reproduce. As Steve Jones puts it in *Almost Like a Whale*, his witty updating of *The Origin of Species*:

> "Evolution is an examination with two papers. To succeed demands a pass in both. The first involves staying alive for long enough to have a chance to breed, while the mark in the second depends on the number of progeny."

Thus, if an individual organism is born with a difference that improves either its ability to stay alive *or* its ability to reproduce, then that individual will pass with higher marks. The progeny of the favoured individual will also possess this difference and pass it on, generation after generation. When the difference is sufficiently large, the old unaltered members of the original population may not be able to compete and will die and become extinct. In this manner, by small but cumulative changes in individual organisms passed on through many generations, do new species emerge and old ones disappear.

Acceptance of Darwin's theory of natural selection came slow and with much opposition, but it is now seldom disputed by biologists.

"How extremely stupid not to have thought of that," remarked T.H. Huxley, on reading *The Origin of Species*.

Its acceptance is due to the simple truth that at present it is the only scientific viewpoint that explains a wide range of apparently disparate facts. Numerous reviews and books, such as the essays of Stephen J. Gould, Steve Jones, Richard Dawkins and many others, document the paleontological, geological and biological findings that support evolutionary theory. Perhaps the most compelling evidence is the growing body of DNA analyses that establish links between a multitude of species that morphologically and microscopically appear to have no commonality. The DNA links support the concept of an evolutionary tree of life, as implied in *The Origin of Species*, that began as a single, simple unicellular organism and eventually, after three billion years of evolution, has spawned all species.

David Attenborough's popular BBC series and book *Life on Earth* (published in 1979) suggests a common origin of life forms by curving the timelines of the major phyla towards convergence at approximately 2–3 billion years ago. Marked dissimilarities between phyla, shown in Figure 4.2, have been used by antagonists of Darwin, such as Hoyle and Wickramsinghe in their book *Evolution from Outer Space*, to hypothesise that some mechanism other than natural selection must have occurred to produce radically different life forms at discrete intervals of time. Seeding of the Earth's atmosphere or seas by fragments of DNA or RNA carried by the dust in the tails of comets is one of the favourite alternative hypotheses. Proof or disproof of such conjectures is not possible, at least at present, so they do not die but remain as grist for the mills of the media.

Various versions of an evolutionary tree of life based upon the DNA equivalence of various components of cells have been constructed. Figure 4.3, modified from Mason's book *Chemical Evolution*, shows an evolutionary tree using the nucleotide sequences of the transfer RNA in the small sub-unit of the ribosome. All

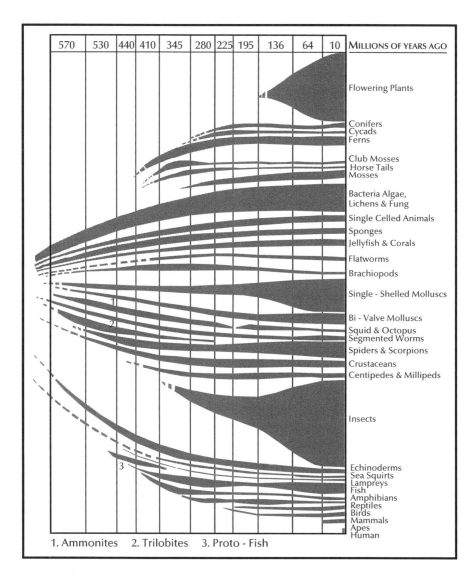

Figure 4.2 Expansion of the number of life forms with time. The width of the columns indicates the relative abundance of the species. (Modified from David Attenborough, *Life on Earth*, Collins/BBC, 1979.)

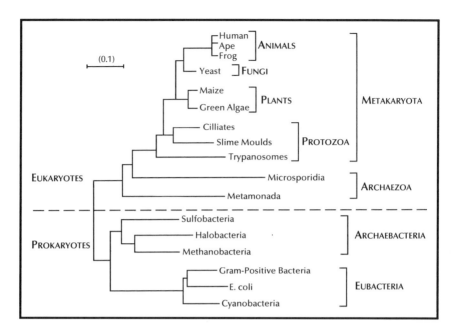

Figure 4.3 Evolutionary tree based on the nucleotide sequences of the rRNA in the small subunit of the ribosome. (Adapted from S.F. Mason, *Chemical Evolution*, Clarendon Press, 1992.)

presently living creatures — and, it is hypothesised, all those in the past — contain organelles in their cells, called ribosomes, which are required to assemble proteins (see Figure 1.4 in Chapter 1).

Ribosomes function by using transfer RNAs to code the assembly of amino acid sequences into proteins. Thus, by comparing the nucleotide base pair sequences of DNA of various organisms, it is possible to determine the extent of their similarity or difference. This is the molecular biology equivalence of determining how far two people live from each other by comparing their telephone numbers. If the first two digits are 44 for two people you know, they both live in England; if the next four digits also match, i.e. are 0207, you know they live in Central London. If these are different, then they live in different towns in Great Britain. In this way you could easily establish the geographic distribution of any number of people throughout the world by knowing their telephone numbers.

With a modest amount of historical research you could use a similar method to map the geographic migration of families by comparing the changes in their telephone numbers over the last century. Such a study would show that some families have stayed clustered together in a village or city for many generations while other families have migrated all over the world. The progeny of the families that have migrated to distant lands are much more likely to have different occupations and different physical appearances than those that stayed put.

We know that the differences in a migratory people are a result of both an altered genetic pool and a different environment. The same pressures for divergence with adaptation have occurred over billions of years for individual organisms regardless of their complexity.

The long term consequence of divergence with adaptation is an evolutionary tree, such as that shown in Figure 4.3. Details of the logic and the 20 years of analytical steps used to construct this type of tree are summarised in *Chemical Evolution* by S.F. Mason.

Life forms exist today in conditions believed to be nearly equivalent to those on the Earth several billion years ago. Archaeobacteria live in extreme environments, such as the super-heated sulphurous vents at the bottom of the sea, or the thermal hot springs and caldera of Yellowstone National Park in the United States. Such species survive in hot water saturated with salts and sulphur compounds at temperatures of 60–95°C. The divergence of these aerobic eukaryotes from anaerobic prokaryotic species appears to have occurred as long as 2.7 billion years ago, based upon findings in 1999 by Australian scientists. The discovery of fatty molecules, called sterols, in ancient shale in northwestern Australia moves the date for emergence of complex life forms backwards from 1.5 billion years ago to as much as 2–2.7 billion years ago. However, it is still unclear whether these primitive life forms had developed other distinguishing characteristics of eukaryotes, such as mitochondria or chloroplasts.

Figure 4.4 Death of modern trees in the Yellowstone National Park, Wyoming, USA, while "primitive" algae survive in the thermal springs' run-off.

There could well have been an intermediate form of life or even myriads of intermediate life forms, during the lengthy period from 2.7 billion years ago when eukaryotes began to flourish and evolve into the countless complex species shown in Figure 4.3.

One of the wonders of the world is that some living relics of these ancient times are still competing favourably with modern life forms. Figure 4.4 shows one result of the competition. The mineral-laden overflow of boiling water from a thermal spring in Yellowstone

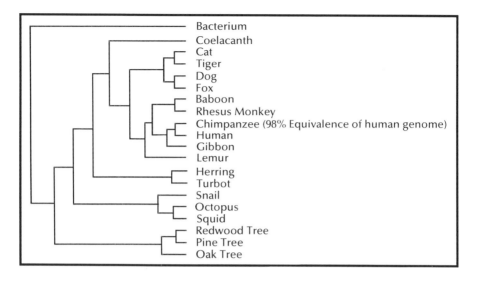

Figure 4.5 R. Dawkins (*The Blind Watchmaker*, Penguin Books, London, 1991) reports, "This family tree is correct. There are 10^{22} other ways of classifying these twenty organisms and all are wrong."

National Park has killed a grove of deciduous trees, but a variety of simpler organisms flourish. Evolution by natural selection selects species to survive based upon chance. When a genetic change enhances survival in a specific environment, the changed organism survives. When the environment is altered, the organism may no longer be favoured and it may die. Thus, the simple organisms in the hot springs thrive but the complex trees die. The ebbs and flows of changes in life forms and environments for countless generations have given rise to the diversity and complexity of the evolutionary tree of life we observe today.

Let us examine two more family trees. Figure 4.5 is a family tree modified from Richard Dawkins' *The Blind Watchmaker*, comparing the relative similarity of 20 familiar life forms based upon protein molecule sequences. Dawkins calculates that there are 8.2×10^{21} other ways of classifying these 20 organisms, but this is the only correct one.

Figure 4.6 Evolutionary tree as seen by Sydney Harris (*Einstein Simplified*, Rutgers University Press, 1989), updated by L.L. Hench).

Figure 4.6 is a marvellous cartoon by Sidney Harris of the evolutionary march of progress. The humour of the latter figure lies in the fact that it pokes fun at one of our most cherished beliefs — that evolution is directed "towards reinforcing a comfortable view of human inevitability and superiority", as discussed in Chapter 1 of *Wonderful Life*, by Stephen J. Gould. "The march of progress is the canonical representative of evolution — the one picture immediately grasped and understood by all."

The contrasts between Figures 4.5 and 4.6 illustrate the issues that underlie the emotional response to the sequence of evolution for many people. Figure 4.5 shows that humans have only very distant similarities to oak trees and bacteria. This is easy to accept.

It is the middle of the family tree — the lines that show chimpanzees to be our close relations — that produces emotional rather than scientific reactions. The fact, determined by DNA analysis, that chimpanzees have a genome equivalence of about 95–98% to humans is very hard for some people to accept. One of the most telling similarities between chimpanzees and humans is our gene for making beta-globin, a critical component in the oxygen delivery system in our blood. Two beta-globins and two alpha-globins make up the haemoglobin in our red blood cells. Robert Shapiro, in *The Human Blueprint*, summarises, "Although our beta-globin chain is the same as the chimp's, it differs from the gorilla, gibbon and rhesus by one, three and eight amino acids, respectively."

It is important that we do not overreact to such scientific conclusions. Shapiro goes on to say, "Individual humans can sometimes show the types of changes that have marked species evolution: duplication or loss of genes. Loss of one copy of the beta-globin gene need not be serious, but loss of both can lead to severe anaemia." Thus, understandably, the 1% variation in the genome of each of us has enormously more practical significance than arguing about the 2–5% variation between us and chimpanzees.

A final, simplified way to chart the increased complexity of life forms over three billion years is to focus on the structural differences in living organisms. These major evolutionary structural modifications are illustrated in Figures 4.7 and 4.8.

For one billion or more years life consisted of single-celled organisms with small segments of DNA isolated from their surroundings by only one membrane. The first significant structural change was incorporation of specialised structures *within* the single cell; the organelles were separated from the rest of the cell by their own membrane. The result of this symbiosis was increased efficiency in energy conversion. Cells that incorporated organelles called mitochondria began the evolutionary pathways that eventually led to humans. Cells that incorporated chloroplasts ultimately led to the diverse range of plant life. All three structural types exist and flourish today as they have for hundreds of millions of years.

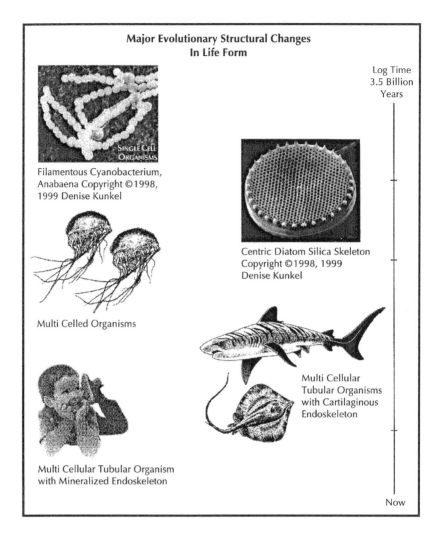

Figure 4.7 Major evolutionary structural changes in life forms over 3.5 billion years.

Cell membranes, composed of a lipid bi-layer and intra-membrane proteins, isolate the metabolism and replication of cells from their environment. But the membrane can be destroyed by mechanical forces, dehydration, radiation, pH and temperature cycles. A second major structural change protected cells from their harsh environment.

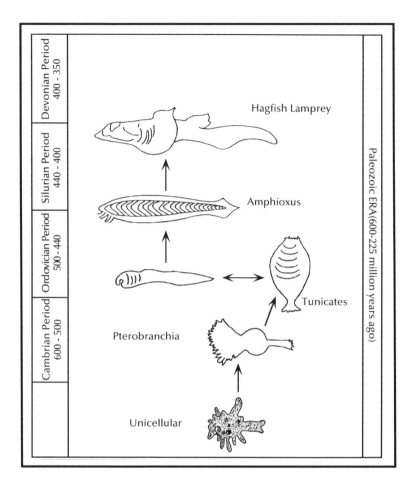

Figure 4.8 Emergence of vertebrates from the larval forms of tulnicate-like creatures. (Modified from Hoyle and Wickramasinghe, *Evolution from Space*, Dent and Sons, London, 1981.)

Mutations of the structure of the membranes made it possible for the cell to encase itself in a protective cocoon, called an exoskeleton. Inorganic elements dissolved into the sea by weathering of rocks were used to build these protective shells. A multitude of species based upon the use of lime (CaO) appeared, such as the stromolites, corals and molluscs.

Another pathway for structural diversity was based upon the use of soluble silicon (Si), the most common element, with oxygen, of the mineral surface of the Earth, to build exoskeletons. The cell wall of diatoms, called a frustule, is composed primarily of hydrated silica and organic constituents. Morphological designs of diatoms are enormously diverse, with more than 20,000 species existing today. Understanding the molecular mechanisms responsible for silicification of diatom frustules is important, because these organisms are responsible for fixing 60% of the nitrogen in the biosphere and are at the bottom of the food chain.

A problem in understanding the mechanism of biosilicification is the apparent absence in all life forms of a metabolic pathway for the element Si. This is an intriguing conundrum, because hydrated Si is present in the connective tissue of all mammals, as discussed in Chapter 5. Thus, the evolutionary structural changes that result in organisms protected by exoskeletons may also be implicated in later structural changes that led to mineralised endoskeletons (see Step 5 in Figure 4.7).

Consequently, several years ago, two of my former research colleagues, Keith Lobel and Jon West, and I decided to explore biosilicification using quantum-mechanical calculations similar to those described in Figure 3.3.

The results are summarised in Figure 4.9. There is a 24-step reaction pathway that incrementally binds silicic acid molecules dissolved in seawater onto the protein template of the diatom. The reaction path is down hill energetically and has only small energy barriers. Thus, the exoskeleton of the diatom can form by photocatalysis without requiring enzymes to be present. This result may account for the fact that a very large number of species, all with distinctive architectures of their silica frustules, have exhibited equal evolutionary survival.

Another major evolutionary step in structural development was the multi-cellular organisation of tubular forms (Figure 4.7). Single cells are limited in size by the diffusion distance of food molecules into the cell and waste by-products out of the cell. When organisms

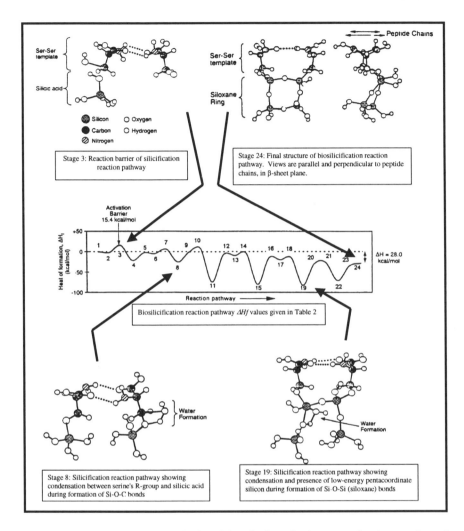

Figure 4.9 Reaction pathway calculated for the biosilicification of a diatom frustule (exoskeleton). (Modified from Lobel, West and Hench, *Marine Biology*, 1996, Vol. 126, pp. 353–360.)

evolved into tubes, they solved the diffusion problem. Food can come in one end and waste out of the other. We are living proof of the versatility of this structural advance. An enormous number of phyla evolved following this revolutionary change in structure.

Tubular organisms with narrow walls are very susceptible to damage. Thus, at some point a further structural change evolved that led to internal reinforcement of the tube. Endoskeletons provided a great improvement in structural stability and enhanced the mobility and competitiveness of the organisms. Sharks and rays today still have cartilage endoskeletons, relics of this early evolutionary advance.

Cartilage endoskeletons work marvellously in a buoyant seawater environment. Occupation of land where the load of gravity must be borne required the final, critical structural evolutionary modification. The collagen produced by the cartilage-growing cells presumably underwent a small mutation, leading to what we now call Type 1 collagen. The unique feature of this special type of collagen is that it can be reinforced by the precipitation of crystals composed of calcium and phosphate ions, called hydroxyapatite (HA). The HA mineral crystals embedded within a collagen matrix create a very tough and strong material — bone. Bone is able to withstand very high mechanical loads and still remain flexible. Bone-mineral-reinforced endoskeletons are the last stages of the evolutionary structural progression that has led to erect species, such as man.

Why is the debate on teaching evolution important? How can it have an impact on ethical decisions of life and death, the objective of this book? I submit that there are at least two topics (which we address in later chapters) that are influenced by evolutionary concepts discussed in this chapter. The first topic we will call "The Importance of Species Similarity". Consider the so-called evolutionary trees of life presented above. The genetic differences between individual humans are less than 0.1%. Chimpanzees differ from humans by about 2–5%; the differences between other mammals and humans are very much larger. These genetic differences must be considered when animals are used for transplants, as discussed in Chapter 6. Only animals that are quite close to the human genome can be used for genetic modification to make immuno-tolerant transplants. Even when a species is genetically similar to humans there are still many uncertainties about their resistance to viruses and prions. Transmission of unknown genetic material from animals to humans is likely to remain a concern for many years to come.

A second concern is the ethical limits of using life forms to benefit humans. It is generally accepted that raising animals for food and clothing is morally acceptable as long as the species are substantially genetically different. Cows, pigs and sheep are acceptable whereas primates are not. However, use of animals of all species for the testing of drugs and medical devices continues to be challenged by groups on moral grounds. It is encouraging that details of the genetic similarities of many species to humans, reviewed in the figures of this chapter, may make it possible to design cell and tissue culture experiments to test the efficacy of some drugs without requiring the use of living animals. This is an important ethical advance.

Bibliography

1. *On the Origin of Species by Means of Natural Selection*, Charles Darwin, John Murray, London, 1859
2. *Almost Like a Whale: The Origin of Species Updated*, Steve Jones, Doubleday, London, 1999
3. *Life on Earth*, David Attenborough, Collins (BBC), 1979
4. *Paradigms Lost: Tackling the Unanswered Mysteries of Modern Science*, John L. Casti, Avon Books, New York, 1990
5. *Wonderful Life: The Burgess Shale and the Nature of History*, Stephen J. Gould, W.W. Norton, New York, 1989
6. *Science and Creationism: A View from the National Academy of Sciences* (2nd edition), National Academy Press, Washington, D.C., 1999
7. *The Wonderful Mistake: Notes of a Biological Watcher*, Lewis Thomas, Oxford University Press, Oxford, 1988
8. *Chemical Evolution*, Stephen F. Mason, Clarendon Press, Oxford, 1992
9. *Evolution from Space*, Fred Hoyle and N.C. Wickramsinghe, J.M. Dent & Sons, London, 1981
10. *The Blind Watchmaker*, Richard Dawkins, Penguin Books, London, 1991
11. *Reconstruction of Cell Evolution: A Periodic System*, Werner Schwemmler, CRC Press, Illinois, 1979
12. *Evolution: An Introduction*, Stephen C. Stearns and Rolf F. Hoekstra, Oxford University Press, Oxford, 2000

5
Ethical Issues of Implants

Introduction

In the previous chapters we discussed the requirements for life and how life may have arisen and evolved from a non-living earth. In this chapter we begin to examine some of the scientific, technical and ethical issues associated with extending life and preserving the quality of life. We have seen that science does not provide all the answers to many fundamental questions. In fact, we learned that the closer a scientific question approaches philosophy and metaphysics, the greater is the uncertainty in the answers. Important consequences of that uncertainty are felt when we discuss implants, transplants and biotechnology, the subjects of this and following chapters. These are not theoretical topics. They affect the life of each of us.

These topics are of concern to all of us because in addition to death and taxes there is another certainty in this life. It is the certainty that we are all wearing out. For most of us some parts will wear out before we die and return to dust. It is probable that every reader knows someone with a hip or knee implant, an intraocular lens, a heart bypass, or a person who wears dentures. These replacement parts, also called prostheses, substitute for living tissues that are damaged or deficient.

It is important to remember that all implants are man-made. They are non-living. However, their purpose is to replace living tissues. We learned from Chapter 3 that we still do not know how living substances originated from non-living materials. We may never know.

However, if we want to maintain a high quality of life, it is necessary to replace worn-out body parts as we grow older, even if we don't have all the answers. It is also necessary to accept that man-made spare parts will not be as good as the living parts they replace. They cannot repair themselves, as can most living tissues and organs. This means that the success of an implant must always be considered from the viewpoint of uncertainty.

Uncertainty means that you must accept that there will be relative success and relative failure when part of your body is replaced. There is a finite probability that any implant will outlast you. There is also a finite probability that the implant will fail before you do. We have no choice but to accept this uncertainty. That is one of the reasons why an approach to life based upon quantum theology, a theology of acceptance of uncertainty, is consistent with many of the problems in today's technological society.

We need to remember that it is only during this century that the length of life has increased to the point where implants have become necessary. Many of the ethical dilemmas we consider in this book arise only because we live in a culture where life has been prolonged from an average of 45 years to an average of 70+ years. The dilemmas that result are new and there are no easy solutions, as we will see.

The Need for Implants

For centuries, when tissues became diseased or damaged a physician had little choice but to remove the offending part, with obvious limitations. Removal of joints, vertebrae, teeth or organs led to only a marginally improved quality of life. However, human survivability seldom exceeded the progressive decrease in the quality of tissues, so the need for replacement parts was small. During the last century the situation changed greatly. The discovery of antiseptics, penicillin and other antibiotics, chemical treatment of water supplies, improved hygiene, and vaccination have all contributed to a major increase in human survivability in developed countries [Figure 5.1(a)]. Life expectancy is now in the range of 70+ years. This increase in survivability, however, means that many people outlive the quality of their connective tissues [shown in the curves in Figure 5.1(b)].

Some 40 years ago, a revolution in medical care began with the successful replacement of tissues. Fortunately, this revolution coincided with the increase in overall human survivability.

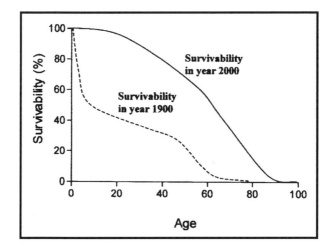

Figure 5.1(a) Increase in human survivability in the USA and Europe during the last century.

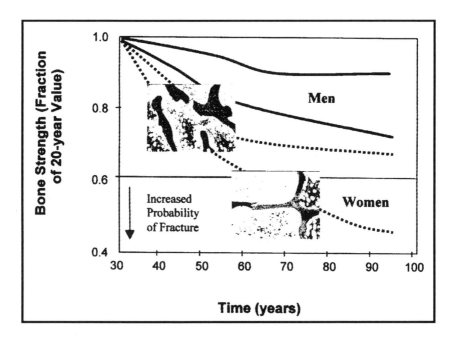

Figure 5.1(b) Decrease in the strength of bone with age for men and women.

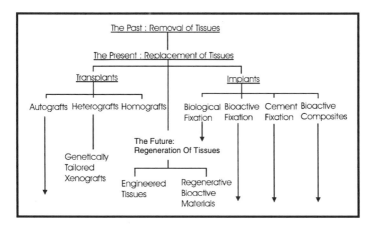

Figure 5.2 The revolutionary changes in the treatment of orthopaedic defects over the last century. [From L.L. Hench, "The challenge of orthopaedic materials", *Current Orthopaedics* **14**, 7–15 (2000).]

Two alternatives became possible (Figure 5.2): (1) transplantation and (2) implantation. Harvesting the patient's tissue from a donor site and transplanting it to a host site, at times even maintaining blood supply, has become the gold standard for many surgical procedures, such as vertebral fusion and coronary bypass.

This type of tissue graft, called an autograft, raises minimal ethical or immunological concerns but does have important limitations. Limited availability, second site morbidity, tendency towards resorption, and sometimes a compromise in biomechanical properties are routine concerns of the surgeon. A partial solution to some of these limitations is the use of transplant tissue from a human donor, a homograft, either as a living transplant (heart, heart–lung, kidney, liver, retina) or from cadavers (freeze-dried bone). Availability, requirement for immunosuppressant drugs, concern for viral or prion contamination, ethical and religious issues limit the use of homografts.

Transplants, both living and non-living, from other species, called heterografts or xenografts, provide a third option for tissue replacement, as illustrated in Figure 5.2. Non-living, chemically treated xenografts are routinely used as heart valve replacements (porcine) and bone substitutes (bovine), but clinical results are less than optimal. Use of living heterografts from genetically modified animals is a controversial step down the ethical slippery slope and consequently is controversial. Chapter 6 is devoted to the scientific and ethical issues associated with transplants of various types.

The second line of attack in the revolution to replace tissues (Figure 5.2) was the development, or in many cases modification, of man-made materials to interface with living, host tissues; for example, implants made from biomaterials. There are many significant advantages of non-living implants over living transplants. They include availability, reproducibility and reliability. Good manufacturing practice, international standards, government regulations and quality assurance testing minimise the probability of mechanical failure of implants. However, most implants in use today continue to suffer from problems of interfacial stability with host tissues, biomechanical mismatch of elastic moduli, production of wear debris and maintenance of a stable blood supply.

In addition, all present day implants lack two of the most critical characteristics of living tissues: (1) ability to self-repair and (2) ability to modify their structure and properties in response to environmental factors such as mechanical load or blood flow.

The consequences of the above-cited limitations are profound. All implants have limited lifetimes. Many years of research and development have led to only marginal improvements in the survivability of orthopaedic, cardiovascular, dental or sensory implants at more than 15 years. For example, efforts to improve lifetimes of orthopaedic prostheses through morphological fixation (large surface areas or fenestrations) or biological fixation (porous ingrowth) have generally not improved survivability over cement fixation of prostheses developed by Sir John Charnley 30 years ago.

During the last decade considerable attention has been directed towards the use of bioactive fixation of implants, where bioactive fixation is defined as: "Interfacial bonding of an implant to tissue by means of formation of a biologically active hydroxyapatite layer on the implant surface". Bioactive implants have been a part of my life for a long time.

The Discovery of Bioactive Glasses

My interest in implants and ethics goes back more than 30 years. It all started with a bus ride. It was the summer of 1967 and I was on the way to a US Army Materials Research Conference in Sagamore, New York. I shared a bus seat with a colonel in the US Army Medical Command who had just returned from being in charge of supply for 20 medical MASH units in Vietnam. He horrified me with stories of the battlefield casualties that were costing thousands of young men amputations due to land mines and high velocity bullets (which shatter the bone from joint to joint). He challenged me, "Why don't young guys like you make as much effort learning to make materials to repair people as you do making materials to destroy them?"

That challenge became especially meaningful to me a few weeks later when I was asked to teach an adult Sunday school class at the First Presbyterian Church in Gainesville, Florida, on the subject "The Ethics of Response".

Remembering the colonel's challenge, I asked a former engineering assistant of mine, Ray Splinter, who was then in medical school, whether there was a problem with implants to replace bone. Ray's answer was, "Yes. The body tries to reject all of the metal or plastic parts that surgeons use in bone repair."

After a series of meetings with Ray, now an eminent maxillo-facial surgeon in San Diego, and two young orthopaedic surgeons, Bill Allen and Ted Greenlee of the University of Florida, Department of Orthopaedics, we submitted a proposal in 1968 to the US Army Medical R&D Command for funding to develop a new type of implant material. We proposed that it should be possible to make a ceramic or glass material that would form a chemical bond to bone. The central idea was simple. Bone contains an inorganic mineral phase, called hydroxyapatite, made up of calcium and phosphate molecules. So I hypothesised that an implant material that contained calcium and phosphate in the right proportions would not be rejected

by living bone. Now, 33 years later, the idea seems so simple as to
be obvious. But back then everyone believed that materials should
be as inert as possible in order to be used in the body. Therefore, all
implant materials used in medicine or dentistry were relatively inert
materials, like stainless steel or silicone rubber.

A year later, in 1969, the first of two minor miracles happened:
the proposal was funded. It was a minor miracle because medical
proposals are never funded with young engineering professors, with
a PhD instead of an MD degree, as principal investigators. It never
happens. But, because of the enormous casualties of the Vietnam
War, the US Army was willing to take a chance on a new idea and
a new investigator.

After the proposal was funded it was necessary to start the work.
That's when a second minor miracle occurred. I designed three
compositions of a glass that contained calcium and phosphate with
just enough silicon dioxide — the major ingredient in most glasses —
to hold the structure together and enough soda to make the glass
melt in our furnace. We melted the glasses and made small implants
of them, which Dr Greenlee inserted into the thighbones of rats.

Six weeks later Ted called and yelled, "Larry, what is that stuff
you gave me?" He was so excited, I thought for sure the glasses had
killed the rats. So, I quickly replied, "Calm down, Ted. They're only
the first tries. There are lots of other compositions I can make."

I'll never forget his answer. He said, "Larry, you don't need to
make any other glasses. The first ones work. Those implants won't
come out. They are bonded to the bone, I've never seen anything
like it before."

He was right. The glass implants did not come out. Nobody had
seen anything like it before. A bond formed that was as strong as
bone. In contrast, control implants of other materials slipped easily
out of the bone because of the scar tissue formed at their interface.
The minor miracle was not only that the very first implants bonded
but also that the composition of those first implants bonded faster
than any implant material tried since. Eighty-five research centres
are now working throughout the world studying the application of

bioactive implants. Bioactive materials are now being used clinically to replace middle ear bones, replace teeth, prevent teeth from being lost from gum disease, repair the spine, augment bone grafts, and stabilise total hip and knee implants. The material, trade-marked as Bioglass®, is currently sold for clinical use in 40 countries. All compositions of bioactive glasses involve hydroxyapatite bone mineral in some manner and are related to the original ones we discovered in 1969. Compositions and reactions to form new bone are shown in Figure 5.3.

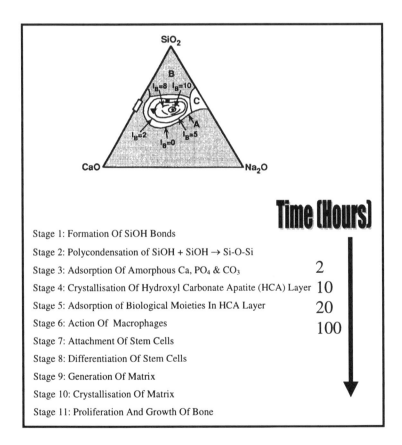

Figure 5.3 Compositions of bioactive glasses and their reactions to regenerate bone. Professor Julia Polak's group at Imperial College has shown that reaction Stages 8–11 are controlled by upregulation of several families of genes.

Working in the field of medical technology has been very satisfying and rewarding. For example, it was how I met my wife and research colleague, Dr June Wilson Hench. June discovered that certain compositions of Bioglass bond to soft tissues as well as bone. She also has helped me immeasurably to survive the struggles of transferring the discovery of Bioglass from laboratory rats into FDA(Food and Drug Administration)-approved clinical use in people.

The First Ethical Issue

Part of the struggle has been a continuing exposure to ethical dilemmas, the subject of this book. My first encounter with commercial ethical issues occurred early in our Bioglass implant programme. I presented our findings at a conference in 1970. Some scientists believed in the results and accepted that a man-made material could bond to bone. Many did not believe that it was possible, or at best were sceptical. Two engineers from a company came to visit in 1972 to check out our findings. We freely discussed everything with them and gave them copies of our reports to the US Army and copies of our research papers.

A couple of years later a friend sent me a copy of a US patent which was issued to those same two engineers and assigned to their company. They had claimed nearly all of our findings as their own, even to the point of copying numerous figures and the text of our first Army report.

In addition, they quickly put the material on the market, calling it Ceravital, a name similar to Bioglass. They had altered it from a clear glass to a crystalline ceramic and changed the composition just a bit. For technical reasons their material did not perform very well. Since they had not done the research they did not understand which variation of the material was best. They used the material in humans before adequate reliability testing was done and this hurt everyone's credibility, including ours.

The consequence of their action that is most difficult to accept, even 30 years later, is that issuance of the patent to the two engineers has hampered for many years our own ability to move the technology into clinical use and help patients. This is because clinical trials and FDA approvals require the financial backing and long term commitment of a corporation. The existence of the patent issued to them frightened off many companies, because they did not want to be involved in a lawsuit. The University of Florida did not have the resources to fight for our rights even though we could prove most of the patent information was taken directly from our reports and belonged to us.

Some university administrators argued that it made no difference who patented and commercialised the discovery. Their position was that a university's role is only to teach and make discoveries. They believed that a university should not be involved in commercialising those discoveries.

This position is highly controversial and requires difficult ethical judgements. The current position of most universities is that they have a moral and financial responsibility to aid in technology transfer and commercialisation. Many universities now promote commercialisation of laboratory discoveries and even encourage the formation of new companies to do so.

My decision was not to get mad, but to get even. Consequently, we started a policy in my laboratory which has continued till today. My policy is to patent all discoveries prior to open publication and disclosure of results to others. My students are asked to assist in preparing the patent disclosures and are always included as co-inventors. Thus, the students learn some of the effort required to make the transition from a concept to a commercial, clinical reality. They share in the experience and share in the royalties, if any result.

Unfortunately, this policy has not prevented business and ethical problems from arising, as we will discuss. However, it has made it possible to see five Bioglass products become approved for clinical sale: a middle ear prosthesis (DouekMED®); a dental implant for maintenance of the jaw-bone for denture wearers (ERMI®); Bioglass powder for restoring bone lost from periodontal disease (Perioglas®); and Bioglass powders for repair of orthopaedic bone defects (Nova-Bone®). Bioglass electrode anchors for extracochlear prostheses for the profound deaf are under test in London. Bioglass powders for use in accelerated wound healing (Dermglas®) are sold extensively in China, and a range of Bioglass products for veterinary applications, pain-desensitising toothpaste and cosmetic applications are in late stages of development.

Number of Implants

With this background let us examine some of the realities and some of the uncertainties associated with the use of implants to enhance the quality of life. First of all, the number of implants used today is staggering. There are nearly three million implants placed each year in the United States. More than 50 different types of implants are used. They are made from more than 40 different materials. Figure 5.4 illustrates the many parts of the body replaced with ceramic implants. Metals and plastic parts are also used for many of these prostheses. For example, there are more than one million intraocular lenses implanted per year. There are nearly a half-million hip and knee implants, several hundred thousand tooth implants, and tens of thousands of bypasses and valves for the heart. There used to be a quarter of a million breast implants. There are implants for fingers, toes, elbows, shoulders and the skull. There are implant powders for cranial and jaw-bone repair and augmentation. The list goes on and on. The numbers grow year by year. And, the need grows even faster.

Where's the problem? With a progressively ageing population, isn't this increased reliance on implants only natural? Isn't it good that there are so many replacement parts available?

The problem is that there are uncertainties in the science and technology of implants, just as there are uncertainties in the big bang, primordial soup and evolution theories. The big difference is that the uncertainties in the origin of the universe or the origin of life do not affect our lives or the lives of our families. The uncertainties in implants, transplants and biotechnology affect every one of us. Health

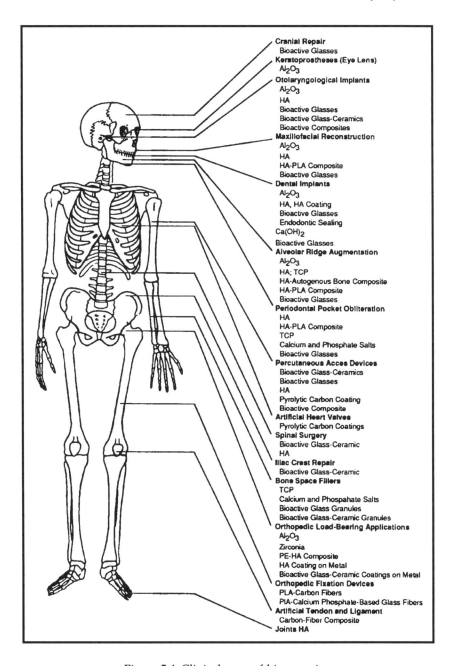

Figure 5.4 Clinical uses of bioceramics.

care costs are higher in the US than in any other country in the world. They are now 12% of its GNP, the most rapidly growing component in the cost of living. However, by many standards the quality of health care throughout all segments of society ranks low on the international scale for developed countries, as we will discuss in Chapter 7.

The health care system in the US creates a very broad distribution of quality of care, which is almost totally dependent on money. This broad distribution of quality often creates ethical dilemmas. The ready availability of implants and the expectation that implants offer miracle cures contribute to these escalating costs.

Implant Failure

Let us examine some specific problems associated with implants. An arthritic or fractured hip is one of worst things that can happen to a person, because it is so painful and it so severely restricts mobility. As people age, especially past 60, the incidence of hip problems increases greatly. A revolution in the treatment of hip disease began in the late 1960's, led by a British surgeon, John Charnley, who was later knighted in recognition of his leadership in the health care of the elderly. His concept was total joint replacement with a low friction prosthesis. He used a metal-on-plastic ball-and-socket joint to substitute for the deteriorated living hip joint. A self-setting polymer, called polymethyl methacrylate, was used to hold the metal and plastic parts rigid in the bone. This procedure is called *cement fixation* of total joint replacements. However, when the polymer sets up in the bone, it fills the space between the implant and the bone rather than cementing the two together, as the name implies.

The big advantage of this method of repair is that the patient can start to use the new hip, albeit very gingerly, the day after surgery. Bone needs to be used in order to be healthy. Consequently, early slow walking is very good for the long term healing of the surgical repair. It is also good for maintaining a good circulation and lung function, as well as being psychologically good for the patient since the pain of walking is gone. There is still pain from the surgery, as total hip replacement is major surgery. However, the pain from the joint is gone. It seems almost a miracle.

Some people describe the repair as a miraculous cure. A Public Broadcasting Service (PBS) programme, *Miracle Materials*, showed a young ballerina performing arabesques and pirouettes after hip replacements. In her interview she described the implants and her recovery as a performer as a miracle, and the interviewer reinforced this interpretation. Such an emphasis on complete restoration of function is in many ways misleading and perhaps even dangerous. The programme omitted several important facts.

The ballerina was physically very fit, with superb muscle tone, unlike most patients requiring implants. She was used to a life of discipline and therefore probably followed her post-surgery instructions meticulously (which is essential for a good result). Most importantly, she weighed probably no more than 100 pounds and was in her thirties. This fact makes a big difference to the success or failure of hip implants. This is because the stress on a hip joint is approximately seven times the body weight. High weight increases the chance of failure. Most hip replacements are in old patients weighing much more than 100 pounds.

When a hip implant fails, it is usually at the boundary between the polymer cement and bone. This is because the cement does not bond to bone. Large mechanical stresses can break down this boundary and cause loosening and movement of the implant in its bony bed. An implant that moves hurts. It also gradually breaks down the bone next to the implant, because that bone is not living anymore. Its blood supply was cut off when the cement was pushed into the bone. The motion breaks down even more bone and eventually the entire implant becomes loose, causes pain and has to come out. Wear of the polyethylene cup produces millions of very small wear particles. The wear debris accelerates deterioration of the bone and failure of the interface.

A failed implant is very bad, for many reasons. The person is older and recovery from major surgery is more difficult. It is often hard to remove the implant and especially the cement inside the shaft of the thigh-bone. The health of the bone has deteriorated due to the presence of the cement and the motion of the implant in the bone. Finally, there is always some chance that the patient will not wake up from the anaesthesia, die from a blood clot, or get infection during the surgery. These complications are all bad. They are also all real. This is the uncertainty we have been talking about. For every implant that is a miracle cure, there is a certain probability that some other implant will be a failure.

The probability of failure ranges from about three to five in one hundred for the first three to five years after an implant to as much as ten to fifteen per one hundred after ten to fifteen years. Figure 5.5 shows the survivability of Charnley type total hip prostheses. The success rate is very high for the first ten to fifteen years, then decreases.

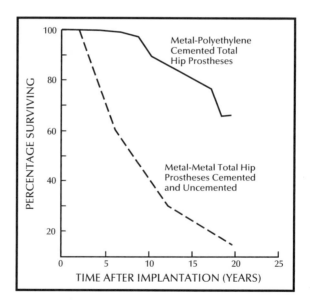

Figure 5.5 Survivability of various types of total hip replacement systems. (Modified from Hench and Wilson, *Clinical Performance of Skeletal Prostheses*, Chapman and Hall, London, 1993.)

The probable life-span of patients who have hip implants ranges from ten to thirty years following the operation. Consequently, there is a fairly high probability, nearly one in three, that a patient will outlive his or her implant. This means that revision surgery and replacement implants constitute an ever-growing portion of the practice of major medical centres. The costs continue to expand as well.

In order to improve the lifetime of implants, especially for younger patients, other methods of anchoring total joint prostheses have been tried. An alternative is called *biological fixation*. In this method the implant is coated with a porous mesh or a layer of porous beads. Bone will eventually grow into the pores. This anchors the implants into bone without use of cement. One of the problems is the length of time for bone to grow into the pores. One approach to speeding up the growth is to fill the pores or coat a roughened surface with hydroxyapatite. The method of *bioactive fixation* with hydroxyapatite is now being used in many surgical centres throughout the world. Ten-year results are good, but it is too early to tell whether this method of fixing implants to bone is superior to cement fixation.

High levels of success for ten years create a problem, however. The problem is that implant companies sometimes move new devices into the clinical market-place with only a few years of animal or human data to support the sales. This is in spite of the fact that the relative success of a new implant design, or method of anchoring, will not be known until ten to fifteen years have elapsed. The patient and the surgeon then face the difficult decision of whether to use one type of implant with a particular level of uncertainty or to use a new type of implant with a relatively unknown level of uncertainty. Thus, the primary uncertainty and question asked by a patient, "Should I have an implant now or wait a few years?", has been expanded to include "Which one of two or three or ten alternative types of implants should be used?".

At present there are few national programmes on implant materials that can provide the scientific basis for answering these questions. There is little research on the long term behaviour of implants for the musculo-skeletal system.

One of the problems is that almost all testing that is done on new biomaterials is done on young healthy animals for short times. However, the implants mostly go into old people with deteriorated bone and must last for a long time. There are almost no long term tests conducted because of the expense and ethical issues of keeping animals in confined test environments for long times. There is little funding for fundamental studies to understand and prevent the breakdown of implants at their interface.

Ten years ago the UK launched an Interdisciplinary Research Centre with long term funding for the purpose of solving these problems. This is a collaborative programme between the School of Engineering at Queen Mary and Westfield College in London and several hospitals associated with the University of London. It involves basic research on cells in contact with materials, bio-mechanics of bioactive composite materials (including Bioglass from our laboratories), and clinical studies using the composites. New bioactive materials have been brought to clinical use from this research, especially a polyethylene–hydroxyapatite composite called Hapex®. This is being used for middle ear prostheses with a high level of success. A solution to the problem of failing joint prostheses, however, has eluded the Centre, as well as all the others working in the field. A new direction for long term survival of replacement body parts is needed. I am convinced that international collaboration rather than international competition is essential for improved health care in the future. The costs to society are too great otherwise. I will discuss these and other new directions in research on implants in a later section.

Ethical Concerns

First, let us return to some of the ethical issues I have observed while working in the field of implants. A long-time friend, Dr Norman Cranin of the Brookdale Hospital in Long Island, New

York, Past President of the Society for Biomaterials, and former Editor of the *Journal of Biomedical Materials Research* and Editor of *Oral Implantology*, is a dental surgeon who has lived through the consequences of reporting failures of dental implants. Twenty-five years or so ago he began to see a number of patients who came to him with problems due to dental implants. In some cases, the patients' dentist, often called an implantologist, who had surgically put the implants into the jaw, did not want to see them when they developed problems. Dr Cranin treated the patients, even though they often had serious complications. He also used dental implants in his practice, as did his residents. The implants were metal blades, posts, or anchors which were used to attach partial or full bridges of artificial teeth and thereby restore the patient's ability to chew. The implants were used for many reasons. In some cases only a few teeth were missing and an implant was used to anchor a partial bridge. In other cases all the teeth were extracted and the implants provided an alternative to dentures. Implants were, and still are, an attractive alternative to dentures. A very large fraction of denture wearers are unhappy with removable dentures. Considering that many millions of teeth are extracted each year and nearly 20 million people in the US are without any teeth, this is a serious problem. Implants appear to be a good solution to the problem.

However, there are difficulties with the implant solution. First of all, the implants are expensive. Some of the systems available today can cost as much as $15,000–$25,000, including surgical costs. Secondly, Dr Cranin discovered that a very large fraction of dental implants were failing. He compiled the data for several thousand cases and reported their relative success and failure in a dental journal. His honesty and his ethical response to the problem led to serious condemnation and professional attacks by some implantologists. His career was threatened. He withstood these attacks and personal and professional abuse, and defended his data. Subsequent studies by other dental investigators found his conclusions to be valid; 25 years ago there was indeed a large chance of failure for many types of dental implants.

Improvements in surgical technique, implant materials, and patient selection and care were implemented throughout the field. This led to a greatly increased success rate of dental implants. The motivation for these improvements is owed to a considerable extent to Dr Cranin's willingness to report failures regardless of the consequences of his actions.

Most of us would agree that his decision and course of behaviour was "right" and therefore ethically and morally correct. Most of us would also agree that those implantologists who put in implants for several thousands of dollars in fees without informing their patients of the high risk of failure, and who then referred them to other dentists when problems occurred, were morally and ethically wrong.

Making such value judgements of relative right or wrong or ethical or unethical behaviour has become more and more difficult. The difficulty is in the expansion of alternatives in today's society. Before implants became available the decision for the dentist or surgeon and the patient was much more straightforward. If a tooth hurt, it came out. If a hip hurt, you quit using it. One problem, one solution. You could question the suffering and, like Job, ask, "Why me, Lord?" But, you had little recourse other than to tolerate the suffering, regardless of your theology or your bank account.

Today's technological society offers many more alternatives, not all of which are equal in terms of risk, benefit and cost. Consequently, in many cases these alternatives produce uncertainties and ethical and moral dilemmas.

For example, what is right or best for you and your family if you are 80 years old, in good health, but are suffering from severe pain due to arthritis of the hip? Should you spend $20,000 for a total hip implant with the risks we have just discussed? Would the $20,000 be better spent on your grandchild's college education? How do the situation and decision change if you are only 70 years old? Is the decision affected if the $20,000 is needed for you to live on for another 10 years? Or, if you have to borrow the money by mortgaging your home? There is obviously no single "right" decision. The uncertainties result in a range of alternatives and a distribution of decisions.

Moral Uncertainties

In the UK these decisions greatly depend on National Health Service (NHS) resources. Long waiting lists for elective total hip replacements have resulted. Several years ago Professor Bill Bonfield, Director of the IRC in Biomedical Materials, showed me a broken femoral stem of a total hip prosthesis. He said, "The patient is 90 years old and this is his fifth failed implant. Should he be implanted with a sixth?" This is an ethical dilemma for the surgeon and the hospital when there are many younger patients still waiting for their first implant. We will discuss these socio-economic issues of health care distribution in Chapter 7.

Professional ethicists and moral philosophers recognise that there are serious problems in resolving the type of moral uncertainties described above. In the introductory chapter of *Contemporary Issues in Bioethics*, Beauchamp and Walters write, "Some moral disagreements may not be resolvable by any of the means discussed. We need not claim that moral disagreements can always be resolved, or even that every rational person must accept the same method for approaching such problems. There is always the possibility of ultimate disagreement."

This admission of the limitations of ethical theories by professional ethicists is both disturbing and satisfying to me.

As I mentioned earlier, many years ago I attempted to lead an adult Sunday school class on the topic "Ethics of Response". Many late evenings were spent reading in an effort to understand various ethical theories. My books included the writings of John Stuart Mill and his successors, who defend an approach called utilitarianism. Their position, that an action is right if it leads to the greatest possible good consequences or least possible bad consequences, seemed completely reasonable to me. Moral rules, thus, are the means to fulfil individual needs and also to achieve broad social goals. A moral life is measured in terms of values and the means to produce the values. This sounds sensible and appears to be similar to the Golden Rule and Christ's Great Commandment to "love your neighbour as yourself". Doing so benefits both yourself and society.

However, the eminent German philosopher Immanuel Kant and his successors, such as W.D. Ross, argue that moral standards exist independently of utilitarian ends. Their ethical theory is termed "deontological", from the Greek word *"deon"*, which means "binding obligation". Thus, deontologists argue that a moral life should not be conceived in terms of means and ends. An act is right not because it is useful, but because it satisfies the demands of some overriding principle of obligation. This appears to be a lot like the teachings of the Ten Commandments and the Mosaic Law. Thus, they maintain that human decisions should be weighed against a higher authority, such as the Revealed Word, the Bible, the Torah or the Koran. Kant stipulates, "One must act to treat every person as an end and never as a means only." This also sounds like Christ's Great Commandment.

However, the differences in ethical theories are fundamental and are difficult, if not impossible, to resolve. Figure 5.6 is my attempt to describe this theoretical ethical dilemma to college students who take my course in biomedical materials and have attended no courses in philosophy. As we saw in Chapter 2, science describes the creation

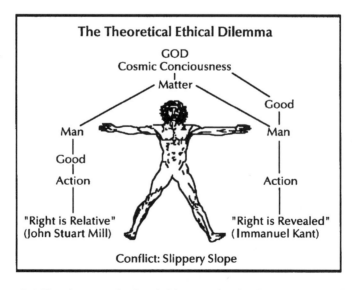

Figure 5.6 The theoretical ethical dilemma that leads to moral conflicts.

of matter from a source that predates man, i.e. matter originates from God, a big bang, a cosmic consciousness, or whatever term is personally satisfying. Man is made from this matter. Moral philosophers who conclude that "right is relative" maintain that the concept of right and wrong and good and evil has evolved from man's social interactions. Thus, what is right depends upon the consequences of action in a specific social context.

In contrast, moral philosophers, such as Kant, who conclude "right is revealed" believe that the concept of good is innate to the individual. Action should be based on the revealed perception of what is right or wrong and therefore is independent of its social context.

This leads to theoretical moral disagreements, referred to by Beauchamp and Walters, that may never be resolved. The conflict between these theories leads to ethical dilemmas. The consequences of these moral uncertainties in the use of transplants, health care distribution, genetic alteration of life as well as birth and death control are the subject of the following chapters. Figure 5.7 illustrates two practical ethical dilemmas that need to be resolved. Our technology-based society has generated infinite desires in people to live a long life. However, there are only finite resources to maintain

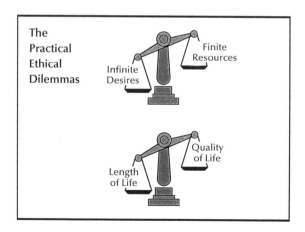

Figure 5.7 The practical ethical dilemmas that lead to personal conflicts.

our quality of life. This imbalance between infinite desires and finite resources creates severe ethical dilemmas.

After reading many books I felt it impossible to decide which theory, or subset of many ethical theories, provided the best basis for making moral decisions. It was during this study and self-examination that I came to reaffirm my childhood faith in the teachings of the Bible. I finally could accept that there was as much guidance in the Bible, in spite of uncertainties, as could be found in ethical theories.

Accepting this was like having a weight lifted off my shoulders. Thirty years later Beauchamp and Walters state, "This problem of how to value or weigh different moral principles remains unresolved in contemporary moral theory." Consequently, we find once again, as in Chapters 1–4, that uncertainty appears to be fundamental to the human condition. In this case it is a fundamental uncertainty in making certain types of moral decisions.

Even at the extremes of behaviour there is often uncertainty. Wanton killing of other humans is condemned as morally wrong in our society and is punished, but killing in war is condoned and even rewarded. Is this rational? Is it moral?

Other situations in our society are equally uncertain as to their moral dimensions. I can cite several examples from my own experiences with biomedical materials. One individual who was president of a company with rights to commercialise our Bioglass implant technology raised many millions of dollars, and brought two products to the market, benefiting thousands. He then stole several millions of dollars from the company and went to a federal penitentiary. Another individual in a different company was also given legal rights to commercialise Bioglass implant technology. He failed to raise any money, did not bring any products to the market, and thereby thousands of patients did not benefit. He succeeded in corporate life in spite of this failure.

A third individual developed a bioactive material, used it for a dental implant, sold his company for several million dollars, and several years later many of the implants failed. A fourth person put a medical material on the market. It succeeded in three clinical applications. However, in a fourth application the material degraded and caused great pain in thousands of patients, leading to corporate bankruptcy and enormous suffering.

Of these four individuals, who was the most morally upright? The precepts of quantum theology say we do not have the basis to judge. The uncertainties are too great. The Revealed Word says we do not have the right to judge. But we do have the right and the responsibility to learn from these examples and the consequences of their failures.

General Moral Principles

There appears to be agreement among moral philosophers regarding three general moral principles that can be used to guide behaviour and make ethical decisions. The moral principles are: respect for autonomy, beneficence, and justice. They are summarised in Figure 5.8.

The principle of respect for autonomy refers to the concept of personal self-governance. It is the opposite of slavery. It assumes that individuals possess an intrinsic value and have the right to determine their own destiny. This principle of a person's right to choose is the fundamental issue that governs most medical situations. Many moral philosophers consider this principle to rank highest in any hierarchy of ethical principles.

The principle of beneficence is that an action or decision should not inflict harm on another, should prevent or remove harm, or promote good to another. Simply speaking, this principle says that it is morally right to aid or to prevent harm to another. Christ's parable of the Good Samaritan is a beautiful illustration of this principle. Conversely, it is morally wrong to harm or to prevent aid to another.

The moral principle of justice requires that like cases be treated alike. This simple concept is called the "formal principle of justice or equality". However, it does not specify *how* to determine equality

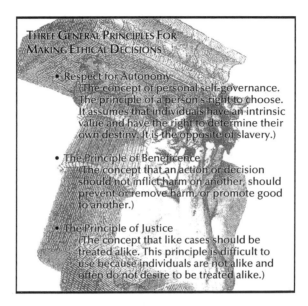

THREE GENERAL PRINCIPLES FOR
MAKING ETHICAL DECISIONS

• Respect for Autonomy
   (The concept of personal self-governance.
   The principle of a person's right to choose.
   It assumes that individuals have an intrinsic
   value and have the right to determine their
   own destiny. It is the opposite of slavery.)

• The Principle of Beneficence
   (The concept that an action or decision
   should not inflict harm on another, should
   prevent or remove harm, or promote good
   to another.)

• The Principle of Justice
   (The concept that like cases should be
   treated alike. This principle is difficult to
   use because individuals are not alike and
   often do not desire to be treated alike.)

Figure 5.8 The three general principles for making ethical decisions.

or proportion in making moral judgements. Therefore, this principle
in its formal sense provides little guidance for decisions regarding
conduct or behaviour. It often contributes to the ethical dilemmas
illustrated in Figure 5.7.

To solve this problem, moral philosophers have developed a
variety of alternatives called material or distributive principles of
justice. Their goal is to provide a basis for judging the relative needs
of people. This is necessary and important because, as we are well
aware, people and their needs are not equal, especially with regard
to health. For example, some people are healthy most of their lives.
Some individuals experience a lifetime of illness or are even born
with genetic defects. Their needs are substantially different, so the
decision as to what is a just distribution of health care is not
straightforward.

Since people are not equal, how can they be treated alike, as the
formal principle of justice requires?

As you might imagine, a variety of principles of distributive justice have been proposed. Some examples have been summarised by Beauchamp and Walters. They include:

Principles of Distributive Justice

(1) To each person an equal share.
(2) To each person according to individual need.
(3) To each person according to acquisition in a free market.
(4) To each person according to individual effort.
(5) To each person according to societal contribution.
(6) To each person according to merit.

PRINCIPLES OF DISTRIBUTIVE JUSTICE

Alternatives

1)	To each person an equal share
2)	To each person according to individual need
3)	To each person according to acquisition in a free market
4)	To each person according to individual effort
5)	To each person according to societal contribution
6)	To each person according to merit
7)	To each person according to age
8)	To each according to status (nobility)
9)	To each person according to gender
10)	To each person according to race

Figure 5.9 Principles of distributive justice.

We can add to this list (Figure 5.9):

(7) To each person according to rank.
(8) To each person according to age.
(9) To each person according to gender.
(10) To each person according to race.

There is a large difference between these alternatives. People will often vigorously defend one of them as just and others as unjust. Consequently, most societies use several of the above principles in combination, in the belief that different rules apply to different situations.

The National Health Service (NHS) in the UK was created to administer health care according to the principles of equal share or equal need. This ideal has become a fiscal impossibility. The growing failure of the NHS to meet the near-infinite demands on it has led to growing criticism. A recent attack from *The Sunday Times* featured the inflammatory headline "NHS: The Bungles, Botches and Blunders". Figure 5.10 quotes *The Sunday Times* version of the failure to achieve justice in British health care.

Many of the difficulties in establishing a national health care policy in the US, which we will discuss in Chapter 7, are due to the enormous disparity in the principles of justice listed above. Advantages and disadvantages of the alternatives for distributing care that have life and death implications are discussed in the following chapters. Ethical conflicts abound.

Some of the problems involved in the widespread use of implants, as described above, are due to the fact they are so readily available. Except for cost, there is little moral concern about the use or distribution of implants. If you want one you can usually have it. The same cannot be said for transplants. Many additional ethical issues are involved, as we will see in the next chapter. However, before we discuss transplants let us summarise some sources of ethical conflict that are created by the difficulty in resolving conflicts between the three great principles of ethics.

NHS: THE BUNGLES, BOTCHES
AND BLUNDERS

*"You arrive for a long-due, urgently
needed operation and find that it is
cancelled. You undergo surgery, then
you're rushed back to hospital just days
after going home. You're hospitalised
for something your GP could have
prevented. You die of an illness that
could have been cured. Things like these
do happen, all the time, across the
country. The personal cost to patients
and their families is obvious. The cost to
the health service is huge. Last month
the National Audit Office reported that
in 1998-99, the NHS faced claims
for clinical negligence totalling £2.4
billion."*

The Sunday Times, 28 May 2000

Figure 5.10 Headline from *The Sunday Times*, 28 May 2000.

Consequences of the Theoretical Problem

Beauchamp and Walters summarise the biggest problem facing ethicists and moral philosophers at the present time thus: "The problem of how to value or weigh different moral principles remains unresolved in contemporary moral theory."

This means that when the Principle of Respect for Autonomy is in conflict with either the Principle of Beneficence or the Principle of Justice, there is no acceptable means of resolving the conflict.

The most important, practical consequence of this theoretical problem is that it leads to uncertainty in assessing an ethical response in individual cases. Guidelines of ethical behaviour can be developed for large population groups, but they may not be accepted by individuals within the group. Individuals often consider general guidelines or restrictions to be unjust if they are excluded. Thus, uncertainty in the relative importance of the three ethical principles leads to conflict between individuals and between individuals and the group.

For example, consider the situation where an implant fails. Conflict may arise between the patient and the surgeon, hospital or manufacturer, or all three. Why is there conflict? The patient chose to have the implant, risks were reviewed and informed consent was obtained; thus the patient's autonomy was respected. The individual case history indicated to the surgeon that the implant and procedure selected had a high probability of success, thereby fulfilling the principle of beneficence. Conflict results, however, when the patient perceives that the principle of justice has been violated. The patient expects not only equal treatment, but also equal results. The patient and his/her family do not care about statistics, and that 85 or 95% of similar cases treated the same way succeeded. They care only that their case has failed.

Source of Conflict

The conflict is due to an unjustified expectation of equal consequences of an act instead of equal performance of the act. The Principle of Justice specifies only that "like cases be treated alike". However, because individuals are different, the results can be different even if the treatment is the same. The difference in results versus expectations can be perceived, wrongly, as unjust (Figure 5.11).

What are the reasons for unjustified expectations of implant success? Three factors, at least, are involved: human nature, technology and greed. It is human nature to want the same things as others. This expectation feeds our market-driven economy. The same is true for implants. People do not desire to live with pain, as they become older. This is reasonable. They learn from the media, their physician or friends that certain implants eliminate pain, and therefore they want an implant if they have painful joints. They do not hear, or are unwilling to accept, that there is a finite risk associated with the

---

SOURCES OF ETHICAL CONFLICT

• Perception that the principle of justice has been violated

• Unjustified expectations of equal consequences of an act instead of equal performance of an act

• Due to emphasis in today's society that technology provides certainty

• Failure to understand that all technological solutions have some risks. A risk-free life is impossible

• Greed feeds on technology and conflict

• Legal system that fails to recognise that all individuals are different

---

Figure 5.11 Several sources of ethical conflict.

surgery and a finite possibility of the failure of an implant. It is human nature to hear only what you want to hear. This results in unjustified expectations and a conclusion of receiving unjust treatment if difficulties arise.

Technology amplifies the problem. New developments in implants are promoted as superior even when long term data for large populations of patients are not available. We live in a technological age where most people want and expect the latest, be it electronics, cars or implants. Along with the latest comes the expectation that the latest is the best. This is often unjustified, but the perception still exists.

Rapid changes in technology also lead to a proliferation of choices. The surgeon and the patient no longer are limited to one decision: Should a hip joint be replaced with a prosthesis? Instead, a series of decisions must be made with regard to: type of stem, type of cup, type of fixation, etc. The statistical basis for risk assessment and beneficence becomes progressively more uncertain the greater the options. The patient may well equate more options with a greater chance of success. This is often false. In fact, the reverse may be the case, i.e. success in a large population decreases as the number of options increases.

Greed can feed on the above factors. As more people want and receive implants, the potential for profits increases proportionally. As more options become available, it is more likely that products will be promoted for the sake of novelty and image rather than for well-established improvements in beneficence.

Economic pressures build to introduce new implant products with only minimal standards of testing in order to have something new to offer. Tests which show that problems may occur are undesirable in this context and therefore are avoided unless required by regulatory pressures. Research to obtain solutions to long term reliability problems is often not done, because to do so is to admit that long term reliability is a problem. Thus, the implant field grows in volume but not necessarily proportionally in beneficence to the larger number of patients. One consequence is an ever-increasing escalation in health care costs, which are discussed in Chapter 8.

Specific Ethical Concerns About Biomaterials

There is a great need to promote testing to avoid long term complications and implant revisions. We need to achieve >85–95% success for implants over 10–20 years. Failure analysis needs to be done for all implants in use or proposed for use, in order to provide a statistical basis for establishing beneficence for the patient. Figure 5.12 illustrates the type of analysis needed.

Clinical results can be classified as those that result in High Beneficence (top curves) or Low Beneficence (bottom curves). Moderate Beneficence lies in between. The ethical Principles of Beneficence requires that an implant meet the high standard of the upper curve because otherwise the principle "first do no harm" may be violated. In other words, any implant that performs in limited trials for 2–3 years, as indicated in the lower curve, should not be

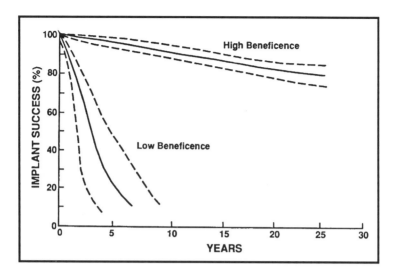

Figure 5.12 Comparison of implant failures as a function of time for high vs. low beneficence to the patients.

put into general use. Any implant that is in general use with results similar to those in the lower curve should be removed from use. Implants of Moderate Beneficence should be subjected to regulatory monitoring to determine the reasons for failures and research funded to improve the performance until the upper curve behaviour is achieved.

Figure 5.12 also illustrates that there exists a distribution of results which broadens with time. Statistical data that can be used to generate such curves need to be compiled by the professional societies for all the prostheses now in clinical use. Patients need to be informed of their expected benefit with respect to this distribution of results. This is one of the few ways to counter false expectations of results.

Also, more controlled center testing of new products needs to be encouraged in order to produce data for plots such as Figure 5.12. A few examples where additional pre-market testing would have been desirable are: intrauterine devices (IUD) for birth control, silicone injections; PTFE powder injections for urinary incontinence; PTFE powder for vocal cord rehabilitation; PTFE/carbon fibre composites for TMJ repair; dense hydroxyapatite cones for edentulous ridge maintenance; plastic and metallic prostheses for ossicular replacement; porous bead coatings on orthopaedic implants; and soya-filled implants for breast augmentation.

Specific Needs to Minimise Ethical Concerns

Clinical results from a number of the implants listed above produce the Low Beneficence curves in Figure 5.12. To avoid this type of performance, we need standard *in vivo* and *in vitro* tests to compare alternative biomaterials under equivalent conditions. Other areas of need to maximise beneficence are:

Long term predictive tests for biomechanical performance;

Predictive *in vitro* tests to determine biochemical factors in tissue response;

Minimisation of animal testing by more effective use of *in vitro* tests;

Avoidance of extensive "me too" development of "new" biomaterials that are only derivative in nature;

Elimination of extensive, repetitive short term animal testing of unloaded, non-functional devices, to be replaced with functional device testing under simulated clinical conditions;

Balancing the requirement of scientists to generate publications and research dollars with the desire to conduct scientific research that can improve long term performance;

Balancing management's often short term outlook with the long term welfare of the patient and society;

Balancing the corporate goal of generation of profit with the need for unbiased product quality.

Summary

Many researchers, clinicians, and manufacturers of implants have been exposed to the consequences of ethical conflicts which can arise when the Principles of Beneficence and Justice are not reconciled. The expectations of the population with respect to implant success will continue to rise. Thus, implants must have increased long term reliability. Failure to ensure this will result in negative consequences for individual patients. Failure will also produce increased governmental regulations and controls. These controls will increase development costs and produce a negative spiral in which fewer manufacturers will be able to afford to produce fewer products and will develop fewer new materials and applications. This negative scenario can be avoided by a concerted effort to improve the long term performance of all types of implants.

The overriding principle "Respect for Autonomy" must be of concern to all of us. An individual depends on information from all sources to make crucial decisions. Such information should always be the best and most complete and be unsullied by commercial or personal preferences. Only by these means can the best decision be made for specific clinical problems.

Bibliography

1. *The Body Shop: Bionic Revolutions in Medicine*, Janice M. Cauwels, C.V. Mosby, St Louis, 1986
2. *Introduction to Bioceramics*, L.L. Hench and J. Wilson (eds.), World Scientific, London and Singapore, 1993
3. *Sol-Gel Silica: Processing, Properties and Technology Transfer*, L.L. Hench, Noyes Publications, 1998
4. *Biomaterials, Medical Devices and Tissue Engineering*, Frederick H. Silver, Chapman & Hall, London, 1994
5. *Biomaterials Science: An Introduction to Materials in Medicine* (2nd Edition), Buddy D. Ratner, Alan S. Hoffman, Frederick J. Schoen and Jack E. Lemons (eds.), Academic Press, New York, 2000
6. *Biomaterials: An Introduction*, Joon B. Park, Plenum Press, 1979

7. *Human Biomaterials Applications*, Donald L. Wise, Debra J. Trantolo and David Altobelli, Humana Press, New Jersey, 1996
8. *First Do No Harm: Wrestling with the New Medicine's Life and Death Dilemmas*, B. Hilton, Abingdon Press, Tennessee, 1991
9. *Contemporary Issues in Bioethics* (3rd Edition), T.L. Beauchamp and L. Walters, Wadsworth Publishing Co., California, 1989
10. *The BMA Guide to Living with Risk*, The British Medical Association, Penguin Books, New York, 1990
11. *What Kind of Life: The Limits of Medical Progress*, D. Callahan, Simon and Schuster, New York, 1990
12. *Journal of the American Medical Association*, May 15, 1991
13. P.S. Saha and S. Saha, *Journal of Long-Term Effects of Medical Implants*, **1[2]**, 127–134 (1991)
14. *Clinical Performance of Skeletal Prostheses*, L.L. Hench and J. Wilson (eds.), Chapman & Hall, London, 1993

6
Ethical Issues of Transplants

Introduction

When your body wears out, there are two routes for repair: use man-made non-living parts (discussed in Chapter 5) or use living transplanted parts (the subject of this chapter). Figure 5.2 in the previous chapter showed that there are three different types of transplants: autograft, homograft and xenograft. There are few ethical issues when autografts are used. Transplanting a vein from a patient's leg in a heart bypass, i.e. which is an autograft, does not require approval from anyone other than the patient and is likely to save his or her life. The great moral principles of respect for autonomy, beneficence and justice are not called into question.

However, replacement of heart, lung, kidney, liver, pancreas, cornea and bone marrow for terminally ill patients can only be satisfied, at present, by transplants from another person, by a homograft. Use of living parts of animals, called xenografts, is generally not acceptable, for reasons which we will discuss later.

The limited availability of transplants gives rise to many major ethical issues. For example, in 1984 a surgical team at the Loma Linda University Medical Center in California implanted a heart from a seven-month-old baboon in a new-born human infant, known as Baby Fae. The xenograft implant stimulated a wave of controversy among the medical community and the public. My wife and co-researcher and I attended a Society for Biomaterials banquet in San Diego a few months after the experiment. The banquet speaker was from the Loma Linda team. Although the audience were experienced medical researchers from the field of implants, the emotional reaction to the use of the baboon heart in a human child was so severe that people left every time the lights dimmed. By the end of the talk the banquet hall was nearly empty.

Many of the concerns regarding transplants are still being debated 16 years later, as described in Caplan and Coelho's book *The Ethics of Organ Transplants*. There is concern about the moral issues of killing animals to get spare parts for humans. Even more disturbing are the rights and wrongs of conceiving a child to provide a bone marrow or other transplant for a family member. The concerns are severe enough that a cover story in *Time* magazine in 1991 discussed the pros and cons of this practice. The practice is still being debated.

The availability and distribution of transplants, regardless of the source, is an important and emotional issue. We often see dramatic cases of children presented on TV with their life or death hanging on the availability of a heart, kidney, liver or lung transplant. The sobbing appeal of the mother and endorsement by a TV or sports personality may result in the necessary organ and money being provided for the surgery — often $100,000 or more.

At the same time thousands of others individuals, of almost surely equal moral merit, die for lack of the organs or lack of money. This certainly violates the moral principle of justice for these individuals. The 1999 Institute of Medicine Report "Organ Procurement and Transplantation" summarises that since the enactment of the National Organ Transplant Act of 1984 the number of people receiving organs in the US has increased annually. In 1998 more than 21,000 people received kidney, liver, heart, lung or other organ transplants. However, on any given day, approximately 62,000 people are waiting for an organ and every day nearly 100 names are added to the national waiting list. Many more are not on the waiting list. Although the number of donors has steadily increased since 1988, donations are not growing as quickly to meet the demand for organs. The Report concludes, "Approximately 4,000 Americans die each year (11 per day) waiting for a solid organ transplant." More than 40,000 persons per year could benefit from kidney transplants. These statistics are summarised in Figure 6.1. This is a fraction of the 80,000 plus patients currently receiving dialysis treatments for kidney

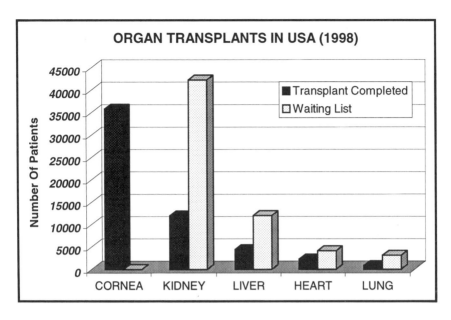

Figure 6.1 Number of patients waiting for organ transplants in the US compared with number receiving organ transplants in 1998.

failure. However, in 1998 less than 15%, or about 12,000 patients, received transplants due to the scarce supply.

Studies indicate that at least 12,000 adults might benefit from heart transplants if a sufficient supply of organs existed. Caplan reports that 7,500 infants have their life threatened with congenital heart disease. Transplants, if hearts were available, could probably save many of them.

In the case of kidney transplants, about one third come from family members who donate one of their two kidneys. However, at present organ transplants are derived primarily from fatal accident victims. There are about 100,000 fatal accidents each year in the US. The system for obtaining and shipping organs from these accident victims is not very efficient. Consequently, only about 20,000 organs are available nationally. This is only a fraction of the number of organs needed, even if the costs are justified and the money is available.

Consequently, there continues to be controversy over the distribution of available organs. There are advocates for all of the principles of distributive justice discussed in Chapter 5. The most dominant force in the US tends to be the principle of distribution according to the ability to pay, either personally or by the local community, moderated by need. To provide a better balance of justice, a lottery system has been widely adopted for allocating organs from fatal accident victims within closest proximity to the recipient. This system is called the United Network for Organ Sharing. The Institute of Medicine Report issued in 1999, cited earlier, discusses many of the issues involved in the distribution of organs, with emphasis on liver transplantation. The issues include:

- Access to transplantation services for low-income populations and racial and ethnic minority groups
- Organ donation rates
- Waiting times for transplantation
- Patient survival rates and organ failure rates leading to retransplantation
- Cost of organ transplantation services

Transplantation of organs from animals is one possible solution to the organ supply problem. This possibility is what led to the Baby Fae experiment. However, this alternative is still highly experimental and subject to much ethical debate. For example, people ask whether our society should promote killing individuals from another species, almost surely primates, to provide organs for humans unless we have made every effort to use the organs of fatal accident victims or terminally ill human patients first.

Others ask, "Is it morally responsible to devote the enormous resources required for any transplant when millions of people lack minimal health care?" Would it not be morally preferable for resources to be allocated to preventing some of the causes of heart kidney and liver disease such as curbing smoking, obesity and alcohol abuse, rather than resorting to transplants? There are no easy answers.

Also, it is important to recognise that the use of transplants commits not only the individual but all of society for one or two generations to the cost of immuno-suppressant drugs. Thus, the approach we take or condone today can mortgage the future for our children and grandchildren.

Issues of Supply

The ethical issues of organ transplants fall into three categories: supply (source and numbers), cost (allocation and availability) and survival (quality and length of life). Our goal in this chapter is to review these issues and determine the extent of the ethical conflict that results. The problem of supply of organs suitable for transplantation stems from the simple fact that the organ must be alive. This requires that the organ come from a donor who is either living or terminally ill but not yet dead. In Chapter 10 we will discuss the issues of death and dying.

Figure 6.1 shows that at the present time the supply of living organs meets only about one third of the demand. With an ever-increasing proportion of the population living even longer life-spans, the supply/demand ratio is likely to drop to one fifth or less. This raises the question of alternative sources of supply. One potential source is the use of anencephalic infants as organ sources. Such infants are born without a forebrain and a cerebrum. This precludes such infants having consciousness. Their organs often still function and could potentially be used as transplants in infants that have congenital kidney, heart or liver disease. A critique of this source of supply (see Chapter 8 in Caplan and Coelho, by Sherman *et al.*) maintains that the number of infants saved by the use of organs from anencephalics would not compensate for "the moral confusion unwittingly introduced into society which would constitute a far greater evil than the good done for the relatively few surviving recipients of these organs".

The American Medical Association Council on Ethical and Judicial Affairs concluded in 1995 that:

"It is ethically permissible to consider the anencephalic neonate as a potential organ donor, although still alive under the current definition of death, only if: (1) the diagnosis of anencephaly is certain and is confirmed by two physicians with special expertise who are not part of the transplant team; (2) the parents of the neonate initiate any discussions about organ retrieval and indicate their desire for retrieval in writing; and (3) there is compliance with the Council's Guidelines for the Transplantation of Organs (see Opinion 2.16, Organ Transplantation Guidelines)."

Numerous alternative sources of organs have been proposed and debated. L.G. Futterman concludes that "the organ supply/demand disparity stems not from a lack of donors but rather from failure to obtain permission to recover viable donor tissue and organs".

Society needs to consider legal changes that make obtaining consent easier. Options debated in Caplan and Coelho include: mandated choice which would require all adults to express written choice of organ donation before death, perhaps as a requirement for obtaining a driver's licence. Since a substantial fraction of organs are derived from motor accidents, such a legal change would go a long way towards solving the organ supply problem. Along similar legal lines, approval of routine salvage organs as a communal policy would eliminate the need for consent.

A further argument in favour of change towards more liberal organ salvage is the difficulty in achieving "informed consent" from the individual or family under tragic circumstances. Beauchamp and Childers discuss the problems and the implications of the concept of "informed consent" in their book *Principles of Medical Ethics*.

Veatch and Pitt argue in Caplan and Coelho against so-called "presumed consent" laws, which they maintain are in effect routine organ-salvaging laws. Their adoption would alter forever the ethical relationship of the individual to society. Routine organ salvaging would involve taking organs without the individual's written consent.

Such laws subordinate the rights of the individual to those of the state. Giving up individual rights to the state, even for the moral good of saving tens of thousands of people, is a profound change in our culture. Most people oppose the loss of such rights.

Purchase of organs continues to be debated. The preponderant opinion is that the rights of autonomy and beneficence favour the present voluntary system of organ donation.

Xenografts as a Solution?

The continuing disparity between the supply and demand of human organs is the reason for a growing interest in transplantation of animal organs. There are two issues: (1) the ethics of breeding and killing animals to save humans and (2) immuno-rejection of the animal organs.

The use of non-human primate organs, such as the baboon heart in the Baby Fae experiment, evokes the greatest ethical concern. Caplan summarises (Chapter 10), "It is one thing to argue that primates ought to have moral standing. It is a very different matter to argue that humans and primates are morally equivalent." Xenografts involving primates can be morally justified on the grounds that, in general, human beings possess capacities and abilities that confer more moral value upon them than do primates. Caplan concludes that the use of xenografts can be morally justified but "The moral obligation to potential recipients would seem to require that systemic farming of animals (for organ transplantation) only be permitted under the most humane circumstances."

The second issue of immuno-tolerance is a severe technical limitation in the use of xenografts. Critical genetic differences between humans require organ transplant patients to use immuno-suppressant drugs on a rigorous daily schedule for their lifetimes. Because of the drugs, they are subject to many ailments, as discussed below. Genetic manipulation of animals will be required to make xenografts feasible. Chapter 7 discusses the ethical issues in making such alterations to the genome.

Cost

The ethical issues associated with the cost of transplants are difficult because they require assessing the unanswerable question "What is the value of a human life?". The answer of course is very different if you are concerned with the intrinsic worth of the life of a single individual, such as yourself or a family member or a friend, than if you are judging the relative worth of one member of a social group of tens of thousands or millions of individuals. Moral philosophers offer little help, as discussed in the last chapter, because they cannot agree on theoretical grounds. For a single person the answer is often "whatever the cost, my life is worth it". For a large social group that answer is impractical.

At present we are going to side-step this conundrum. We will return to it in Chapters 8–10. For the remainder of this chapter we assume that every life is worth saving by a transplant and examine the personal and social consequences of doing so. Discussion of the cost of transplants usually emphasises the economic issues, such as the cost of surgery and the length of hospitalisation. A liver or heart transplant costs $250,000, and a kidney transplant costs $90,000 over a five-year span.

These costs are indeed large, but in proportion to the population they represent only a small fraction of the costs of health care. Deciding whether this level of expense to save the lives of a small group of individuals is at present a community decision based upon resource allocation, discussed in Chapter 8. The moral decision is whether or not to recruit a transplant team for a community hospital and equip the necessary facility. Once the transplant unit is in place there is no longer an ethical issue as to whether or not to use it.

The ethical concern becomes how to ration it. Seldom are there sufficient organs to provide transplants for everyone in the community who could benefit.

Emphasising transplant economics misses a second, and perhaps even more compelling, cost of transplantation: the cost to the recipient. Cost is more than money; it is also pain and loss of quality of life. It can be loss of freedom. The ethical debate of present day transplants often overlooks the fact that most recipients will require the use of immuno-suppressant drugs for the rest of their lives. The drugs are costly. They must also be taken on a precise schedule if they are to be effective. The cost of a slip-up can be fatal. The cost of living with the knowledge that your new heart, lung, kidney or liver may deteriorate if you make a mistake is an enormous burden to a person.

The downside of the drugs that suppress immune rejection of the transplant is the fact that a recipient is susceptible to every infection that comes along. There is no such thing as the "common cold" to the transplant patient. Every virus may be fatal. Who can measure the cost of being held hostage by your own body?

Survivability

The third critical issue in the ethics of transplants is survivability of the patient. Is the projected lifetime of a transplant recipient worth the enormous personal and social costs? There is no clear answer. Data from the Scientific Registry of the US United Network for Organ Sharing (UNOS) show for a four-year period (1987–1991) that first-year survival of 4,830 heart transplant patients was 82%. That means one in five patients survived less than one year. In the same period 86 patients with failing transplants were re-transplanted. Only 57% survived one year. Five-year survivability of heart transplant patients is less than 50%.

Data on patient survival following liver transplantation show similar results to heart transplants. During 1987–1991, 8,539 patients in the US received the first liver transplants through the UNOS system and 76% lived for at least one year. Patients who were already on life support had a much lower survivability of only 60%. When the transplants failed, second or third transplants were made and one-year survival dropped from 50 to 35%.

Figure 6.2 summarises survivability of University of Pittsburgh liver transplant patients up to three-and-a-half years. Patients who were relatively low risk had a high one-year survival rate (91%)

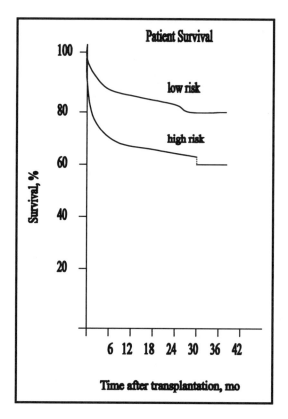

Figure 6.2 Patient survival following primary liver transplantation in adults stratified as to being low risk or high risk. (Data from Bronsther *et al.*, in Chapter 26 of Caplan and Coehlo.)

which was maintained at 80% after 42 months. In contrast, only 71% of the very sick patients who had only a few days to live due to terminal liver failure survived for one year after a liver transplant.

Matching the antigens of donor organs to the recipient is one way of improving patient survival. Gaston *et al.* (Caplan and Coelho, p. 310) report that 1,004 kidney transplants that had a zero-antigen mismatch had one-year survivability of 88%. This is a 10% improvement over the 79% survivability of 22,188 recipients of mismatched kidneys in the same period.

These data illustrate several important points:

First, you have an about 80% chance of living for more than a year after most transplants and an about 50/50 chance of living for more than five years.

Second, if your first transplant fails, the probability of surviving a second or third transplant is very low indeed.

Third, the sicker you are, the less likely it is that you will survive for very long with a transplant.

The Institute of Medicine Report "Organ Procurement and Transplantation" leads to several additional conclusions:

Fourth, organ allocation areas need to be matched with patient populations using statistical analyses.

Fifth, rates of pretransplantation and transplantation mortality are more meaningful indicators of equitable access than waiting times.

Sixth, policies related to transplantation access must be consistent with the medically acceptable cold storage times for organs: liver, 12 hours; pancreas, 17 hours; kidney, 24 hours; heart, 4 hours; lung, 6–8 hours.

Another Institute of Medicine Report, "Approaching Death: Improving Care at the End of Life", leads to another conclusion:

Seventh, death is sometimes preferable to the consequences of having a transplant.

Conclusions

The issues of transplant availability, cost and survivability should provide the basis for making a moral judgement as to whether transplantation is a desirable option to prolong a life. When a patient is beyond a certain age, or level of illness, the ethical principles of beneficence and justice indicate that subjecting him or her to the trauma and subsequent personal costs of a transplant is unwise.

Transplantation of a second or third organ cannot be justified on moral grounds, even though the consequence of not doing so is death of the patient. In Chapter 10 we discuss the complex ethical issues of death. One of our conclusions is that "avoiding death at all costs" is immoral. This conclusion is especially relevant in making decisions regarding transplants. Just because it is possible to do a transplant does not necessarily mean that one ought to do a transplant. Social and surgical guidelines to establish this boundary need to be very tight. The respect for autonomy for the terminally ill patient should not bias counsel for a transplant towards a decision of "life at all costs".

This point is emphasised dramatically in a poignant review of the field of transplant medicine by Renee Fox and Judith Swozey, authors of *The Courage to Fail: A Social View of Organ Transplants and Dialysis* and also *Spare Parts: Organ Replacement in American Society*. They have decided to leave the transplant field and in doing so state, "By our own leave-taking we are intentionally separating ourselves from what we believe has become an overly zealous medical and societal commitment to the endless perpetuation of life and to repairing and rebuilding people through organ replacement — and from the human suffering and the social, cultural and spiritual harm we believe such unexamined excess can, and already has, brought in its wake."

Theirs is a strong expression of ethical concern that should not be ignored.

Increasing the length of life without assuring the quality of life is immoral. It is time to examine whether for some people that boundary has been crossed.

Bibliography

1. *Ethics in Medicine*, Milton D. Heifetz, Prometheus Books, New York, 1988
2. *Principles of Biomedical Ethics* (4th Edition), Tom L. Beauchamp and James F. Childers, Oxford University Press, Oxford, 1992
3. *Critical Reviews in Biomedical Engineering*, John R. Bourne (ed.), Vol. 25, Issue 2, 1997

4. *The Ethics of Organ Transplants: The Current Debate*, Arthur L. Caplan and Daniel H. Coehlo, Prometheus Books, New York, 1998
5. "Organ Procurement and Transplantation", Committee on Organ Procurement and Transplantation Policy, Institute of Medicine, National Academy Press, Washington, D.C., 1999
6. "Approaching Death: Improving Care at the End of Life", Committee on Care at the End of Life, M.J. Feld and C.K. Cassell (eds.), Institute of Medicine, National Academy Press, Washington, D.C., 1997

7
Ethical Issues of Genetic Manipulation of Life

A potential solution to the need for transplants is to use genetic engineering to solve the problem at the source, within the living body itself. This is only one of many possible applications of the relatively new field called biotechnology. It is apparent from Indira Vasil's remarkable book *Biotechnology: Science, Education and Commercialisation* that there are many definitions of biotechnology. For this discussion I will use one of the broadest definitions: "Biotechnology involves the production and application of purposefully manipulated biological organisms." Applications can be in many directions, including improvements in the yields of grain crops, fisheries, cattle, chickens or milk production; improvements in creating more efficient vaccines, disease-resistant vegetables, genetically engineered pesticides, etc. Perhaps the world's greatest hope for meeting the food requirements of an ever-growing world population is the use of these advances. An important example is genetically modified rice with extra vitamin A. This GM rice can potentially curb a vitamin A deficiency, which causes blindness in 500,000 malnourished children every year. Greenpeace has announced that its activists will not disrupt or destroy field trials of the crop known as "golden rice" (because of its colour).

As important as improvements in food production is the use of biotechnology in health care — the primary emphasis of this book. As discussed in previous chapters, implants or transplants are often poor substitutes for human body parts. It would be far preferable to use biotechnology, which includes genetic engineering, to stimulate or induce self-repair. Many species, such as salamanders and newts, have this capability to a much greater extent than man, as discussed in detail by P.A. Tsonis in *Limb Regeneration*. The information for repair and regeneration is stored in our genetic code, which is shared with these species. All we need to do is decipher or unlock the code and learn how to control it. In doing so, it is likely we can accomplish something even more useful, i.e. learn how to diagnose and eventually prevent genetic diseases. There is certainly hope for this exciting possibility.

McKusik, from Johns Hopkins University, summarised in 1991 that, of the 1,800 human genes that were mapped to specific sites at that time, disease-producing mutations were known to occur at more than 400. As the map of the human genome becomes ever closer to completion, probably by 2002, knowledge of the number of mutant genes responsible for defects in humans increases almost daily.

It is now known that mutant genes on the 23 pairs of human chromosomes are the most likely to be associated with genetic disease. The altered genes responsible for many serious abnormalities, including Huntington's chorea, Duchenne and Becker muscular dystrophy, type II diabetes and cystic fibrosis, are among the genetic diseases identified so far. Genes on chromosome 1 are associated with Alzheimer's disease and prostate cancer. The X chromosome has been found to be particularly susceptible to disease mutations.

Therapeutic applications using gene splicing are already being used experimentally in humans. This offers the hope of combating incurable genetic diseases such as ADA deficiency, which shuts down the immune system. If this treatment is successful the implications are enormous. Between 1 and 5% of the infants born in the US are afflicted with genetic diseases that have no effective means of treatment. When treatments are available they are usually expensive and are needed for a lifetime, often with serious side effects.

Gene therapy aims to provide the body with healthy replacement genes that can provide the function of the defective gene. In effect, as Dr R. French Anderson of the NIH Clinical Center in Bethseda, Maryland, a pioneering advocate of gene therapy, said in *Time* magazine several years ago, "You can engineer a patient's cells to pump out anything now given by injection, growth factor, factor VIII, insulin, etc. The advantage is that it's a one-time treatment."

Other potential uses of gene therapy include treatment of melanoma, a deadly skin cancer that affects 28,000 Americans annually (including one of my sons). Dr Steven Rosenberg of the NIH, who is a colleague of Dr Anderson, also was quoted as saying in *Time* magazine, "We now use radiation, chemotherapy and surgery, all external forces, on cancer patients. Gene therapy uses the body's own internal mechanism. We're trying to make the body itself reject the disease."

Rosenberg uses a gene that codes for the tumour necrosis factor, a naturally occurring compound that attacks cancer cells. This cancer-fighting gene is spliced onto a carrier virus, called a retrovirus, which inserts itself, with its piggyback fighter, into immune cells extracted from tumours of the melanoma patient. The new fighter gene is built into the cell's genetic material, converting the cell into a guided missile to attack cancer cells. The genetically engineered fighter cells are injected back into the bloodstream of the melanoma patients, where they seek out and destroy the cancer cells. Although early results were encouraging, relatively few new treatments have been produced so far. The promise is certainly there. Future applications will include "attacks on haemophilia, diabetes, Parkinson's disease, Aids, and muscular dystrophy", according to Inder Verma, a leading genetic engineering scientist at the Salk Institute, in the November 1990 issue of *Scientific American*.

The Human Genome Project

Researchers in the US, the UK and Japan began in 1990 the largest project in the history of biological science. The purpose of the project is to provide the ultimate database for detecting genetic defects and designing gene therapies. It is called the Human Genome Project. The goal of the project is to map and then identify all the nucleotide base pairs of DNA in the human genome. The enormity of this goal makes the translation of the Rosetta Stone or the Dead Sea Scrolls seem like solving the *Sunday Times* crossword puzzle. However, only a little more than a decade later, the end is in sight. The first draft was published in 2000, The complete text will follow progressively over the next two to three years.

As we discussed in Chapter 3, all cells known today that reproduce do so by replication. All information required for replication as well as for maintenance or metabolism of the cell is contained within its DNA. The information is stored as base pairs, A–T and G–C, which bind together the strands of the double helix. Each base pair is a single bit of information, like 0 or 1 in a computer, or a letter of the alphabet or a note of music. Combinations of base pairs result in the equivalent of words, sentences, paragraphs, stories and chapters that provide the instructions to produce proteins that do the work. A collection of paragraphs constitutes an instruction manual for a specific function of the cell, such as making insulin or breaking down sugar or fat, thereby releasing stored energy, which the cell uses. Each instruction manual is called a gene.

More complex organisms require more instruction manuals for repair and reproduction. A very simple bacterium, such as *E. coli*, which lives by the billions in our gut, contains 4.6 million base pairs. Yeasts contain about 15 million base pairs. The parasite that causes malaria, a disease that annually affects 200–400 million people, has 25 million base pairs. The human genome is estimated to contain 3,000 million (3 billion) base pairs. The enormous magnitude of information in the human genome compared with simpler organisms is illustrated in Figure 7.1.

The huge amount of information contained within each of the thousands of millions of cells in each human is packaged as approximately 40,000 genes or instruction manuals. No one knows for sure the exact number. It is believed to be 26,000–40,000 genes, according to Dr Craig Venter, from Celera Corp., one of the pioneers in the Human Genome Project.

The amount of 40,000 instruction manuals, even as small folded molecules, is bulky. Consequently, they are packaged together like a cellular handyman's encyclopaedia, called chromosomes. Simple

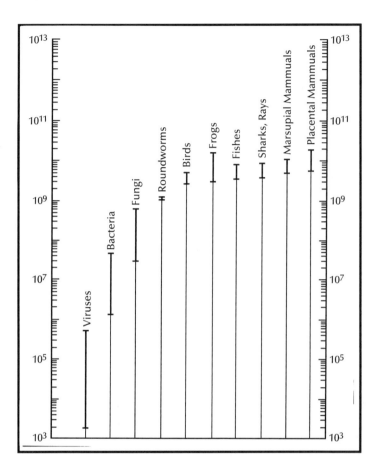

Figure 7.1 The range of nucleotide content in major taxonomic categories. (Based upon B. John and K.R. Lewis, *Chromosome Hierarchy*, Clarendon Press, London, 1975.)

organisms require only the equivalent of one or a few encyclopaedias. Human cells each contain 23 pairs of encyclopaedias, or chromosomes, or 24 if you count the X and the Y as separate, and a 25th if you include the unique DNA information stored in the mitochondrion, the energy-producing unit in our cells.

If you printed out the information in the human genome, using one letter per nucleotide base pair, it would result in 25 sets of encyclopaedias, each set on average half the size of the *Encyclopaedia Britannica*. Not all chromosomes are of the same size, however; some sets will be large and some small. That is a stack of books taller than most rooms.

At present, more than 90% of the instruction manuals or human genes are known and mapped. The number mapped is uncertain due to the proprietary claims of a US firm (Celeron Inc.) which is deciphering the genome as a commercial enterprise. This is a concern that we will discuss later. Some genes are very small and some are very large, as discussed by Matt Ridley in *Genome*. The Human Genome Project identifies the base pair sequence of the gene and then maps its location on a specific chromosome or chromosome region. The result is like knowing which of the 25 sets of encyclo-paedias contains a specific instruction manual and where it is located within the encyclopaedia.

One of the mysteries, to return to our theme of identifying scientific uncertainties and discussing their consequences, is that only a fraction of the DNA in the human genome is estimated to be in the form of genes. Genes tend to be located at the very ends of our chromosomes and, according to Dr Cantor, at one time head of the Human Genome Project at the University of California at Berkeley, "Why this occurs is simply unknown." Dr Venter says researchers are startled to find so few genes in the human genome. The original estimate was 100,000–150,000 genes.

The origin and reason for much of the other 90% of the DNA sequences is not known. It may consist of the scrapbooks or old, out-of-date operational manuals for cellular functions from the dawn of life that are no longer used. Perhaps this material is like the Egyptian hieroglyphics on the Rosetta Stone — useful and, in fact, a dominant language and storehouse of human information 6,000 years ago, but lost and forgotten by all but a handful of scholars today. Without the two, more modern languages also on the Rosetta Stone, Coptic and Greek, we still would not understand the several thousand years of Egyptian history.

Thus, by analogy, the history of life from some 3,000 million years ago may be stored within each of us. Our DNA may contain the relics of the most primitive life forms and the complete story of the origin and evolution of life. We may just not know how to read the language.

Another musical metaphor may help one to comprehend this genetic uncertainty. A sequence of base pairs in a gene may serve the same function in a living cell as a series of notes in a musical composition, i.e. the base pair sequence encodes the synthesis of a specific protein in the cell; the sequence of notes instructs a pianist which keys to strike, a flautist which holes to cover, a violinist which strings to bow or pluck, or a trumpeter which valves to push. The sound made is different for every instrument even though the notes are the same. Likewise, transcription of the same base pair sequence by other organisms may or may not lead to the same protein that is produced in a human cell.

Although there may be solo parts in a musical score, most notes are not played in isolation. They are played in concert with an assembly of instruments that are often playing very different notes. For great music the complementary nature of the notes results in complex harmonies and rhythms that are pleasing. A large measure of the pleasure is the complexity. Highly simplified arrangements of notes, such as in so-called rap music, are not satisfying to ears used to symphonies and concerti, and vice versa. Even the notes of a Schubert sonata played poorly and out of sequence do not sound pleasing. Purposeful disharmonies constructed in some modern music often sound alien.

Genetic mutations may also produce the equivalent of musical disharmonies. If the change affects a vital part of the body, such as the occurrence of sickle cells instead of normal red blood cells, the result is often fatal. A goal of genetic engineering is to identify the site of such abnormalities on the chromosomes and then eliminate the defect. A genetic engineer needs to serve as both composer and conductor to get the music right.

The Genetic Genie

The potential use of information on the magnitude of the Human Genome Project for good is enormous. At the Biovision Conference in Lyon, France, February 8, 2001, Dr Francis Collins, Director of the National Genome Research Institute in the US, predicted that by 2010 "we will have uncovered the hereditary contribution to the bulk of the most common diseases in our society, with predictive tests for perhaps a dozen diseases and interventions to reduce risk".

By 2020, Dr Collins predicted that doctors will be armed with gene-based designer drugs to treat almost every disease. "By 2030," he said, "if all goes well, there should be the kind of comprehensive genomics-based health care where we have individualized preventative medicine based on your individualized risk instead of a 'one size fits all' approach."

However, complex ethical issues also have to be considered, since there is always potential for misuse and harm or injustice as well. Here's a list of questions posed by Dr Inder Verma, professor of molecular biology and virology at the Salk Institute, in his *Scientific American* article on gene therapy a decade ago: "Should gene therapy be applied to improve one's offspring, not only to prevent an inherited disease? Who should be empowered to decide? Is society willing to risk introducing changes into the gene pool that may ultimately prove detrimental to the species? Do we have the right to tamper with human evolution?"

We can add many other questions, such as: If you diagnose an embryo as carrying a defective gene that will result in a lifetime of pain and suffering, is abortion justified?

What if a disease, such as Huntington's chorea, does not hit until 40 or 50 years of age? Who has the right to decide — parents, child or society?

Do you want to know if you have the potential for a serious disease? What do you do about it if you do know? Should life and health insurance companies have the right to know? Who should control the ownership of your genome?

Hearings of the House of Commons Science and Technology Select Committee on February 8, 2001, highlighted one of the above questions. Dr Ian Gibson, MP, a member of the committee, accused a prominent insurance company of trying to set up a "genetic ghetto". He called on the Government to stem the tide of genetic testing by insurance companies.

Self-regulation is clearly not working. These companies are attempting to identify a genetic underclass, which can only lead to profit for them and discrimination against individuals.

The test for Huntingdon's disease is at present the only genetic test that has been passed for use in the UK by the independent Genetics and Insurance Committee.

Half of the states in the US have partial or complete bans on the use of genetic tests for underwriting risks. Austria, Denmark, the Netherlands, Norway and France all impose legal restrictions.

A serious risk is that individuals will be discouraged from taking tests with potential medical benefits because they fear insurance companies will obtain the information and discriminate against them.

Respect for autonomy, beneficence and justice are difficult to apply to the above questions because of the high degree of uncertainty of the consequences of an action.

However, in spite of the uncertainty we must begin to explore these issues. They will not go away. They will become even more critical when the Human Genome Project is completed.

We must also face the reality that for a long time we will learn much more from the Human Genome Project than can be used medically. Frustration and even anger can result. The results will often be difficult to understand and use. Consider the similarity to buying a fancy VCR that can be programmed to record several programmes while you're on vacation. You are given an instruction manual, similar to a gene, and all you have to do is follow it and the VCR will work, as the salesman (the genetic researcher) has convinced you.

However, as most readers know, errors often happen; one button is pushed in the wrong sequence or isn't pushed, or the power goes off and that programme is lost, or you put in a 30-minute tape by mistake or you have forgotten to rewind the tape. The point is that having an instruction manual does not necessarily mean it can or will be followed accurately. What if you order the VCR from a discount catalogue or from the Internet and the instruction manual that arrives is in Japanese or Korean? You probably can't follow it at all. Many of the DNA sequences in the human genome presently appear to be equally indecipherable. Perhaps there is a hidden language. The so-called junk DNA, which is a large part of our DNA, may not be junk at all.

In any case, there is plenty of uncertainty in each of us. It is important that we accept the uncertainty of our own biology, the same as we have to accept the uncertainty of the theories of the big bang, the primordial soup, and evolution. Accepting means we cannot predict all the consequences of life. However, accepting uncertainty does not mean ignoring the challenges. Growth comes from exploration, in spite of uncertainty. Faith proves the assurance that even with uncertainty, life has meaning.

Two issues of uncertainty, regarding genetic manipulation of life, presently dominate public perception. The first genetic genie that is out of the bottle is human cloning. The second is genetically modified (GM) food. Daily articles appear in the media discussing the havoc that these two genetic genies can wreak if the government or the courts of law do not put them back in the bottle. Some responses mimic the furore caused by the discussions on evolution summarised in Chapter 4.

The origin of inflammatory comments regarding genetic engineering is often similar — an uncompromising belief in the sacredness of life, a belief that life created by God should not be manipulated by man. This is a belief that altering the "natural" way of life is violating God's will, as proclaimed in the Scriptures.

The fact that millions of people world-wide espouse these beliefs does not mean that they are true. It also does not mean that they are false. Such beliefs must be respected and considered when the ethical issues of genetic manipulation of life are debated. Let us consider the debate on cloning.

Few topics today generate as much fear or loathing as human cloning. This fear is based in part on reality and in part on superstition. The reality is that there is presently great uncertainty about the outcome of the cloning of human cells. There is also a considerable lack of knowledge about the scientific issues associated with cloning. Uncertainty combined with a lack of understanding generates fear. The fear is amplified by superstition. Fictional accounts of the manipulation of life, from Mary Shelley's *Frankenstein* of 1818 to Aldous Huxley's *Brave New World* of the 1950s and Ira Levin's *The Boys from Brazil* of the 1970s, to countless science fiction movies of the 1980s and '90s, have filled our social consciousness with the belief that the result is evil.

Headlines from respected national newspapers illustrate the difficulty in discussing the scientific and ethical issues of human cloning in a rational manner. Many attitudes are emotional rather than factual.

Headlines

Exclusive: First Cloned Human Embryo: The picture that marks another leap forward in science — but is sure to fuel the moral debate: Fears that a baby could be cloned
> — *Daily Mail*, 17 June 1999

Babies born of science into a moral maze
> — *The Sunday Times*, 26 December 1999

Science ready to let men have babies
> — *The Sunday Times*, 21 February 1999

Having disabled babies will be "sin", says scientist
> — *The Sunday Times*, 4 July 1999

Human cloning hits a national barrier
> — *The Sunday Times*, 26 December 1999

Dr Dolly plans the clones that will save lives
— *The Sunday Times*, 16 May 1999

Dolly is mutton dressed as lamb
— *The Times*, 27 May 1999

Human cloning — a risk too far?
— *The Times*, 11 January 2000

"Immortal genes" found by science
— *The Sunday Times*, 4 July 1999

Stuff nature. Now even Mummy's playing God
— *The Observer*, 26 September 1999

Imperfect children left to die
— *The Sunday Times*, 21 February 1999

Hope for childless as scientists grow lab womb
— *The Sunday Times*, 3 October 1999

Did this man achieve his dream of breeding a master race?
— *The Sunday Times*, 26 December 1999

Playing God: making carbon copy humans
— *The Sunday Times*, 11 March 2001

The Issues

One of the first issues to recognise is that cloning of human cells needs to be divided into two categories: reproductive cloning and non-reproductive cloning, also called therapeutic cloning. The purpose of reproductive cloning is to produce another human being with the same genes as the donor cell. The objective of therapeutic cloning is to grow tissues and organs to repair a human being suffering from disease or trauma.

Dolly, the cloned sheep, proved that reproductive cloning of mammals is possible. However, Wilmut, Campbell and Tudge, the developers, emphasise in their chronicle of Dolly, *The Second Creation*, that Dolly was the *only* success out of 277 reconstructed embryos. They conclude, "All the results so far suggest that, by present techniques, foetuses produced by nuclear transfer are ten times more likely to die *in utero* than foetuses produced by normal sexual means, while cloned offspring are three times more likely to die soon after birth. Deformities also occur. These setbacks are distressing for people who work with animals — and surely mean that any immediate extension into human medicine is unthinkable."

In addition to the scientific issue of the reliability of cloning, there is also concern about the biological age of the clone. Wilmut *et al.* report, "Dolly's genes come from a six-year-old animal, and although Dolly after two years looks like a normal two-year-old, we have yet to see how her life will pan out. The telomeres on her chromosomes are reduced in length. The shortening of the telomeres does not cause ageing, it merely reflects ageing, but when the telomeres have disappeared altogether, the cell does die."

They go on to conclude, "There is no reason to suppose that Dolly's telomeres have shortened enough, or will do so, to affect her lifespan but we cannot foresee how a longer-lived animal, such as a human being, would be affected."

Raanan Gillon, Professor of Medical Ethics, Imperial College School of Medicine, comes to a similar conclusion in his Stevens Lecture for the Royal Society of Medicine, "Human Reproductive Cloning — A Look at the Arguments Against It and a Rejection of Most of Them." He says, "Given the limited potential benefits of reproductive human cloning, the benefit/harm analysis does not seem at present to create much moral pressure to argue against it."

Gillon refutes emotional prohibitions on various moral grounds and concludes, "All the arguments for a permanent ban on human reproductive cloning fail and most of the arguments for even a temporary ban fail. However, four arguments in favour of a temporary ban do currently succeed.

"The first is that at present the technique for human reproduction by cloning is simply not safe enough to be carried out in human beings.

"The second related argument is that, given these safety considerations, the benefits, including respect for the autonomy of prospective parents and the scientists who would assist them, are at present insufficient to outweigh the harms.

"The third is the argument from distributive justice [see Chapter 5], but this is only insufficient to prescribe a low priority for state funding for human reproduction cloning.

"Fourth, respect for autonomy within a democratic society requires adequate social debate before decisions are democratically made about socially highly contentious issues... ."

Gillon is especially concerned about governmental prohibitions that affect human autonomy and our freedom to make our own decisions. Directives from organisations such as the General Assembly of the World Health Organisation, which proclaim human reproductive cloning as "ethically unacceptable and contrary to human integrity and morality", and the positions taken by the European Commission, the Council of Europe and UNESCO, which essentially forbid cloning, have dangerous overtones.

Gillon maintains, "A central issue that underlies the cloning debate, and indeed the overall debate about the new genetics, and is of deep moral importance, is the need to protect the genetic underpinning of human autonomy and free will.

"The underlying moral challenge is our shared obligation to protect ourselves against the predations of the control freaks, whether they are the control freaks of state or religion or science or big business, or simply crooked gangsters who seek to use us for their own ends."

The plans of Dr Severino Antinori in Italy to clone a human baby and the proclaimed intent of the Raelians religious cult in the US to recreate a 10-month-old baby boy who had died in a hospital accident are proof that these ethical issues must be debated urgently and openly.

Unfortunately, the fear of human cloning may extend to the use of embryonic cells for growing tissues and organs for repair or replacement of diseased, damaged or aged parts of the body, i.e. therapeutic cloning. The concept of taking cells from a patient or a donor and then modifying them by genetic manipulation, if needed, followed by expanding their number in cell culture so they can be injected or implanted into a patient, can potentially revolutionise the treatment of many illnesses. Notably, treatment of disorders of the neural system, such as Parkinson's or Alzheimer's disease, may be possible by this method.

The new fields of genetic engineering and tissue engineering offer the potential of circumventing many of the limitations of implant or transplant technology discussed in Chapters 5 and 6. It is critical that these fields proceed towards clinical and commercial reality to solve many of the problems of an ageing population. Failure to do so is immoral. Research in these fields can now move forwards.

Approval of Stem Cell Research

In the UK a landmark decision was made on December 20, 2000, by the House of Commons and confirmed on January 23, 2001, by the House of Lords, to amend the 1990 Human Fertilization and Embryology Act and permit stem cell research on human embryos up to 14 days old for the purpose of developing treatments for chronic human diseases. Reproductive cloning remains illegal.

The decision followed the recommendations of an expert panel chaired by Professor Liam Donaldson, the Chief Medical Officer, and was supported by the British Medical Association, who contacted all 659 Members of Parliament, urging them to back "stem cell" research.

Patient-support groups, research charities and scientists all pushed hard for the research. They argued that stem cells extracted from embryos could be used to treat brain diseases such as Parkinson's and Alzheimer's, repair damaged organs and cure diabetes.

The scientific foundation for the decision is that embryonic stem cells are unprogrammed cells which can become any type of tissue.

Prior to the amendment, research using human embryos was allowed only for strictly limited purposes. Hundreds of thousands of surplus embryos, created by IVF (*in vitro* fertilisation) treatment for infertile women, are destroyed.

During the debate leading up to the vote, pro-life groups insisted that the technique, which would involve cloning human embryos up to 14 days old, is morally wrong and sets a dangerous precedent.

Some members compared the research to Nazism and the creation of Frankenstein's monster. Others claimed it would lead to cloning of humans.

However, Yvette Cooper, the Public Health Minister, assured the House that stem cell research did not represent a "slippery slope" to human cloning, which would remain illegal. She said the research could hold the "key to healing within the human body", giving hope not only to those suffering from degenerative diseases but also to cancer and heart disease victims. She added, "There are immense potential benefits from allowing this research to go ahead particularly for those suffering from dreadful chronic disease."

Conservative Ann Winterton, the leading opponent of the move, said it was a "cruel hoax" to claim that voting against the regulations was tantamount to depriving the sick of a cure. She said, "It is both untrue and unspeakably cruel to tell families who suffer from genetic disease and other problems that a vote against cloning is a vote against providing them with any hope for the future. It is the view of the expert group that the long term promise of stem cells from adult tissue could equal or surpass that of embryonic stem cells. There is no evidence other than wishful thinking that research on human clones can kick-start that process."

Liam Fox, the Shadow Health Secretary, said he was morally against the use of embryo cells and had not been convinced there was no alternative. He added it was unrealistic to think such research could be halted; so tough rules were needed to set the moral boundaries.

A hearfelt plea to approve the change came from Labour Aberdeen South MP Anne Begg, who is confined to a wheelchair by a rare degenerative disease. Stem cell research had "enormous potential" for people suffering from a wide range of conditions and scientific and medical opinion was almost unanimous in its support, she said.

Mrs Begg said, "Almost everyone who suffers from a degenerative disease is desperate for this research to go ahead, including many for whom the results of the research will come too late."

Merits of Stem Cell Research

There are several characteristics of embryonic stem (ES) cells that make them especially attractive for developing therapeutic treatments. First, ES cells are pluripotent. They are isolated from the developing blastocyst, which at the earliest stages of development consists of no more than a ball of cells. This can be grown *in vitro*. All the cells are equivalent, with no distinguishing characteristics: under appropriate conditions the ES cells will change (called differentiation) into all the specialised cells of the human body, i.e. into specialised cells of nerves, skin, bone, heart etc. (Figure 7.2). It is this pluripotential promise that provides hope for treatment of disorders of the nervous system where repair is not normally possible. For repair of skeletal defects, adult mesenchymal stem cells can be used for therapeutic purposes, as argued in the House of Commons debate. However, as few as one cell in 100,000 in the bone marrow stroma are likely to be stem cells. Moreover, the number of stem cells able to proliferate and differentiate into productive bone and cartilage cells decreases considerably with age.

A second, and related, feature of ES cells vital to their therapeutic use is the fact that they can divide readily for many generations with their numbers increasing exponentially. Therefore, only a relatively few ES cells are needed to start a therapeutic treatment; the second, third and fourth generations can be subdivided indefinitely to seed onto tissue engineering scaffolds or be injected into a treatment site.

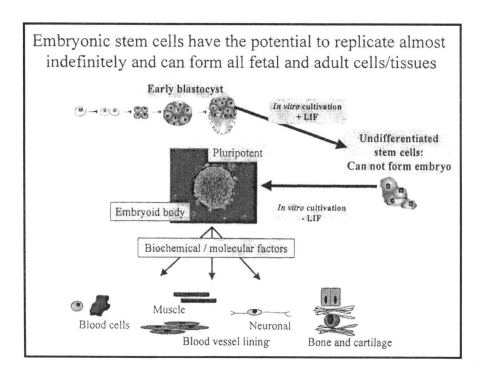

Figure 7.2 Potential use of embryonic stem cells to replicate tissues and organs using tissue engineering technology. (Figure from Professor Julia Polak, Director of the Imperial College Tissue Engineering Centre, London.)

A third important feature of ES cells is that they are more amenable to genetic manipulation than adult cells, which are already partially or fully differentiated. This genetic tractability of ES cells means that they can be altered to correct genetic defects prior to their therapeutic use. Genetic alteration to render the cells immunotolerant is one of their most important advantages. As discussed in Chapter 6, a severe limitation of life-saving organ transplants is a lifetime's dependency on immuno-suppressant drugs. Growing organs *in vitro* from ES cells that are rendered immunotolerant would provide the ultimate solution to this problem and eliminate the shortage of donors.

We are presently a long way from growth of a heart, lungs or joints in the laboratory, let alone an organ production line. However, the first steps towards such an end point have been taken. ES cells have been successfully differentiated into the ectoderm, mesoderm and endoderm at the Tissue Engineering Centre at Imperial College by Dr Anne Bishop and Professor Julia Polak, and into bone cells by Dr Lee Buttery and Professor Polak.

Other teams in the US and Europe have produced from ES cells controlled cell cultures of haemopoetic precursors, neural cells, adipocytes, muscle cells and cartilage cells. In addition, cardio-myocytes derived from mouse ES cells have been shown to form stable cardiac grafts in the living animal, and tissue-engineered neural cells have been used to reconnect nerves.

Translation of these findings into commercial and clinical treatment still has numerous hurdles to overcome. Long-term stability of engineered tissues and organs is still uncertain. All natural structures in the body grow under complex combinations of biochemical stresses and in a 3D environment dictated by neighbouring tissues. All tissues develop a source of nutrition and connections to the nervous system as they grow. None of these vital features have yet been achieved in the engineering of tissues or organs. Even repair of skin or cartilage — relatively simple tissue — with tissue-engineered products presently offers only limited success.

Ten years ago, when this book was started, none of the above concepts were even considered as feasible alternatives to prolonging life and increasing the quality of life. Now, with the completion of the Human Genome Project and the associated gene chip technology, it is possible to work systematically towards a new generation of health care technology. Uncertainties and ethical issues still exist, as discussed above. We have also seen that informed governments can and will address these issues and progress will continue. The fear of uncertainty has not prevailed. For the present, the genetic genie has been judged to be good. However, it is a good idea to keep an ethics eye on the genie for many years to come.

Bibliography

1. *Biotechnology: Science, Education and Commercialisation*, Indra K. Vasil (ed.), Elsevier Science Publishing Co., 1990
2. *The Body Shop: Bionic Revolutions in Medicine*, Janice M. Cauwels, C.V. Mosby, Missouri, 1986
3. *The Human Blueprint: The Race to Unlock the Secrets of Our Genetic Script*, Robert Shapiro, St. Martin's Press, New York, 1991
4. *Controlling Technology: Contemporary Issues*, William B. Thompson (ed.), Prometheus Books, New York, 1991
5. *Journal of the American Medical Association*, May 15, 1991
6. *Fortune*, July 1, 1991
7. *Animal Experimentation: The Moral Issues*, Robert M. Baird and Stuart E. Rosenbaum (eds.), Prometheus Books, New York, 1991
8. *Ethics in Medicine*, Milton D. Heifetz, Prometheus Books, New York, 1996
9. *Critical Reviews in Biomedical Engineering*, John R. Bourne (ed.), Vol. 25, Issue 2, 1997
10. *The Ethics of Organ Transplants: The Current Debate*, Arthur L. Caplan and Daniel H. Coelho, Prometheus Books, New York, 1998
11. *The Second Creation*, Ian Wilmut, Keith Campbell and Colin Tudge, Headline, 2000
12. "Human reproductive cloning — a look at the arguments against it and a rejection of most of them", Raanan Gillon, *Journal of the Royal Society of Medicine*, Vol. 92, January 1999
13. *Limb Regeneration*, Panagiotis A. Tsonis, Cambridge University Press, 1996
14. *Genome: The Autobiography of a Species*, Matt Ridley, Fourth Estate, London, 1999
15. *The Sequence: Inside the Race for the Human Genome*, Kevin Davies, Weidenfield, 2001

8

Effects of Technology on Health Care and Resource Allocation

In Chapters 5–7 we saw that the technological developments of implants, transplants and genetic engineering have greatly expanded the alternatives available for treating deterioration of the human body. We saw that these developments are expensive. They also have a degree of risk or uncertainty associated with them. The costs and risks frequently lead to ethical dilemmas for a family. Costs incurred in health care in the US often result in decisions that can ruin families financially and emotionally. Cost limitations in the National Health Service (NHS) in the UK often lead to delays in treatment and at times fatalities. In Chapters 9 and 10 the ethical issues associated with the ability to control technologically the beginning of life and the ending of life are discussed. The goal of this chapter is to examine the technology and ethical issues involved in the availability of health care to maintain and ensure the quality of life. Our concern is the ethics of resource allocation.

We discussed in Chapter 5 that moral philosophers do not agree which principles of material distribution are most just. The US has evolved a hodgepodge health care system which tends to operate primarily on the principle of allocation by ability to pay. A consequence is the terrible statistic that more than 40 million US citizens have no health care insurance.

The Americans without health care insurance are generally the poor, the elderly and the minorities, African Americans and Hispanics, who constitute the largest proportion of the unemployed or underpaid work force. Because they cannot pay, the quality and availability of their treatment are often less than for the remainder of the population.

In the UK the NHS has the goal of providing universal health care. This laudable objective is under continuous erosion. Long waiting lists for elective surgery, depleted hospital staff, unavailability of hospital beds and loss of morale in overworked and underpaid physicians and nurses are daily news items in the UK.

Inequity or injustice in distribution appears to be only a symptom of the real problem — which is rising costs. Health care costs in the US are growing like a tumour. In 1950, $13 billion was spent on health care in the US. That was less than $100 per person. This figure tripled in 15 years. By 1965 health care costs were $42 billion while the population had increased by only a few per cent. The effect was a jump in costs to $205 per person. In the 15 more years to 1980, costs skyrocketed to $250 billion, an increase of six times. The costs per person escalated to more than $1,000. In the following ten years they tripled again to $650 billion. That was over $2,500 per person, or $10,000 for a family of four. By this year 2001, as Daniel Callahan predicted in his important book *What Kind of Life*, health care costs in the US will have exceeded $1.5 trillion.

This runaway inflation in health care costs needs to be compared with the US gross domestic product (GDP) in order to eliminate general economic inflationary factors. Anyone who goes to replace a 15-year-old car knows that it is going to cost at least twice what the same model used to cost.

However, the increase in health care costs greatly exceeds inflation. In 1965, health care consumed about 6% of the GDP. By 1990 it had doubled and now accounts for 12% of the GDP. By the end of this year it will be up to 15%. Thus, the escalating increase in the costs of health care is far more than general inflation.

Perhaps an even better indicator of the cost problem in US health care is a comparison with major economic trading partners. The difference is shocking. In 1960 the US expended 5.2% of its GDP on health care. The others were close: Canada, 5.5%; Great Britain, 3.9%; Germany, 4.7%; Japan, only 3.0%. However, by 1986 our costs had more than doubled to 11.1%, substantially higher than any of the others. No other country was even close to 10%. Great Britain had held costs to 6.2%, Germany 8.1%, Canada 8.5% and Japan 6.7%. During the same time period, as Callahan points out, the percentage of GDP devoted to education in the US decreased from 6.4% in 1973 to 6.2% in 1986. Defence spending has also remained constant or decreased in terms of inflationary dollars.

You might assume that US citizens are getting more and better health care than those countries since they are spending 50% more for it. If you do, you are wrong, at least for the country as a whole. In nearly all categories the US ranks behind, not ahead of, most of the other developed countries. For example, in the '90's it ranked 14th in infant mortality, 15th in life expectancy, and down at 27th in heart attack recovery. Not a very impressive record, is it?

The level of satisfaction with the health care system sank to a dismal low of 10%, as discussed in a May 15, 1991, editorial in the *Journal of the American Medical Association* (*JAMA*). Ratings of health care in the US have not improved much in the decade since. In contrast, Canada and Germany reported satisfaction ratings of 56 and 41% ratings respectively. From the same source, support for a universal health insurance system of some sort received an overwhelming 92% by labour union leaders, 72% by the public and 67% by corporate executives. These high figures resulted even if it meant increasing taxes.

So, there is general agreement that something is wrong with the current health care system in the US and there is a growing consensus that something should be done about it.

Similar concerns are expressed about the NHS in Great Britain. John Le Fanu, in his perceptive book *The Rise and Fall of Modern Medicine*, describes the magnitude of the problem and analyses some of the causes. A continuing rise in expectations of the health care system is one of the major factors.

A goal of this chapter is to discuss the causes of the problem of rising health care costs and the growing inequality and injustice in distribution. A second goal is to suggest what can be done about it. Rather than giving specific literature citations in this discussion, I want to recommend five excellent books that have been useful in analysing this problem. Each offers a very different but important perspective.

First, the British Medical Association's *BMA Guide to Living with Risk* (Penguin Books, 1990) provides the statistical base as well as discussions on various risk factors associated with living in our modern technological world.

Additional details are presented in two reports issued by the Royal Society in Great Britain — *Risk Analysis, Perception and Management* (1992) and *Science, Policy and Risk* (1997).

Second, *Intensive Care*, by Thomas Raffin, Joel Shurkin and Wharton Sinkler (W.H. Freeman and Co., 1989), provides a highly personal feel for the drama that occurs daily in the thousands of intensive care units around the country. This little book makes it clear why the expense of saving a life "at all costs" is so high. It also makes it quite clear that regardless of the extent of money and love devoted to patients in intensive care units, two or three of every ten patients will never leave them alive.

Third, Bruce Hilton's book, *First Do No Harm: Wrestling with the New Medicine's Life and Death Dilemmas* (Abingdon Press, 1991), strips away the jargon and rhetoric from dozens of bioethics issues and presents a down-to-earth, sometimes sarcastic, but easy-to-understand analysis. An example: "Surgery to close bed sores is something we can budget for. It challenges the best in medicine. Having a nurse's aid drop by to turn the patient, so bedsores won't develop in the first place, is not nearly as interesting. It doesn't keep the wheels of medical commerce spinning, either."

Fourth, *To Err Is Human: Building a Safer Health System*, published in 2000 by the National Academy Press, describes the importance of human error in medicine and outlines potential ways to improve the health care system.

Finally, for the big picture and a challenge to change, to emphasise caring as well as curing, and for a plea to re-establish a balance in our health care system, you should turn to Daniel Callahan's book *What Kind of Life: The Limits of Medical Progress* (Simon and Schuster, 1990). One example (p. 252) to whet your appetite: "...one reason we come to want indefinite progress is that we constantly upgrade our needs to stay one jump ahead of our achievements; the more we get, the more we want. We manage to keep ourselves dissatisfied, no matter how much better off we become. This is an old story in human affairs. We have added something new to it, however. There is the fear of ageing and death in the company of modern medicine; the greater sense of illness and vulnerability that progress ironically instils in us; the growing inability to find a way of coming to grips with the reality of death, a reality now seemingly transformed into a wrenching choice rather than a deliverance of fate; and the anxiety occasioned by our capacity to transform our biological condition without a comparable capacity to transform our social condition."

Powerful words and very true.

To live at all costs and to maintain life at all costs has become the standard of our society. This is both the blessing and the curse of today's technology. So, to the task: "What are the causes of the problem and what can be done about them?"

As in any complex situation involving large numbers of people, there is no single cause. However, I suggest that there are a few primary factors responsible for the problem. These are:

(1) An expectation that medical technology can keep us alive as long as we want;
(2) A legal system that often requires prolonging life past the point that is desirable either personally or societally;
(3) An economic system that perpetuates and expands both the first and the second factor;
(4) Greed that feeds on all of the above.
(5) Human error that magnifies factors 1–4.

I Deserve a Long Life

Let's examine these four factors and see how they reinforce each other and lead to runaway inflation of costs and unfair distribution of resources. The first factor is the expectation that all of us are due our three-score-and-ten years of life, as indicated in the Bible. If the average is 70 or 75 years, then that is what we want and feel we deserve, no less. More is OK but less is not.

No longer is 50 or 60 years of life acceptable, as was the case early in this century when 45 years was the average life-span. Any time greater was considered a bonus. It is written in verse 10 of Psalm 90, "The days of our years are three score and ten; and if by reason of strength they be four score years, yet is their strength labour and sorrow; for it is soon cut off, and we fly away." Many ethical and economic problems arise from a false expectation of everyone having a right to an old age. We address some of these problems in the following chapters on birth control and death control.

I emphasise that the right to live to an old age is a false expectation. Historically there has always existed a broad distribution of life-spans for humans.

For life in general, for the millions of species on this planet, there is invariably a broad distribution of life expectancy. We have seen scores of documentaries illustrating in dramatic fashion the relatively few survivors from the thousands of fertilised eggs that are laid or deposited by thousands of different life forms. We have witnessed on TV how the predators of land, air and sea continually select and weed out the young, the weak, the sick and the old of the herds, flocks and schools of animals, birds and fish.

This is the natural order. Trial and error, success and failure, good genes passed on, poor genes lost, generation after generation.

This is the basis of natural selection and adaptation. This is the basis of life as we know it. This is humankind's legacy as depicted in the Bible, in myth, and in science.

The central feature and focus of this legacy is that with life comes uncertainty. If you are lucky you will be born healthy, strong or clever and will survive. If you are unlucky you will be weak, sick or stupid and will die. If you are lucky your good traits will be passed on to your children and your children's children. If you are unlucky your poor traits will eventually die out. However, there is never certainty. Sickly children have been born in the past to healthy parents and still are today. "From clogs to clogs in three generations" is an old English expression which illustrates that the legacy of wisdom, diligence and prudence is as unpredictable as the English weather.

In short, with life comes risk. With life comes uncertainty.

Our generation now questions this natural legacy of life. We want to prevent risk. We want to eliminate uncertainty. We want to prolong life regardless of the costs.

Why has this attitude arisen? It is largely because so many more of us are living longer. And, we want to continue living even longer. However, this attitude and goal of society is based on a false understanding of the reasons so many of us are living longer. The real reason is a shift in the statistics of mortality. Chapter 1 of *Living with Risk* points out that the increase in life expectancy experienced during the past century has largely been due to decreasing infant mortality. "Among children and young people the death rate is now less than one tenth of what it was in the 1840's, while it has been halved for those in their sixties and reduced by around one third for the oldest groups."

Only about 2% of deaths now occur between birth and age 14, and accidents represent a large portion of these. Consequently, many more people born today survive until they are 70 than ever before. The BMA book says, "A corollary of this is that the more people can be saved from dying from premature death, the more people will die from what we used to call 'old age'." It goes on to say, "The care of the aged, and the management of the hazards to which they are particularly susceptible (such as falls and extremes of temperature), will be a major challenge in future years." This prediction has now come true in most of the developed world.

The enormous change in mortality rates during the past century has largely been due to the elimination of infectious disease as a cause of death, especially tuberculosis and death in childbirth, either of the infant or the mother. This enormous shift in the causes of death is illustrated in Figure 8.1. The consequences are shown in Figure 8.2. There is a very large shift in population towards older people.

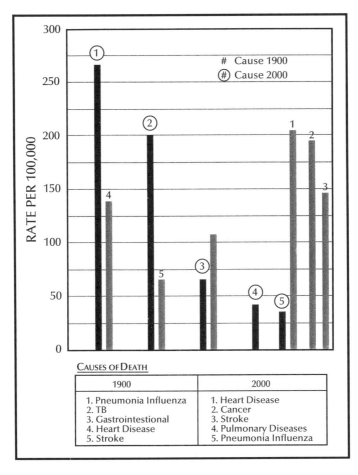

Figure 8.1 Primary causes of death in the USA in 1900 compared with the year 2000.

In the 1900s the leading causes of death were pneumonia, influenza, tuberculosis and gastrointestinal infections. Now, the three main killers are: first, diseases of the blood circulation system (including heart attacks and stroke), which are responsible for nearly one-half of deaths; second, cancer (about one in four); and third, accidents, murders or suicides (including violent injury and poisoning). These three causes account for three-quarters of all deaths. Death of young men is now largely due to accidents and violence.

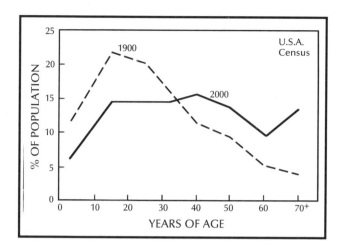

Figure 8.2 Shift in the age of the population in the USA during the last century.

Only after 40 does heart disease take over as the major cause of death among all males.

An important statistic is that no single disease — other than diseases of the circulatory and the respiratory system, or cancer — is responsible for more than 2% of all deaths. An equally important statistic is that the single most important factor in increasing death is smoking. The second is alcohol abuse. The third is obesity. The effectiveness of all three primary killers is greatly increased by smoking and alcohol abuse.

If the goal of our health care policy were primarily to prolong life, then the single most effective allocation of resources would be to eliminate smoking. The second would be to minimise excessive alcohol consumption. However, this is not an easily accepted conclusion. Tobacco lobbies have been very effective in thwarting legislative changes despite enormous social pressure. Enhanced public awareness of the dangers of smoking or of breathing smoke-filled air has led to the enforcement of smoking bans in airplanes and public buildings, and the creation of no-smoking areas in most restaurants. Taxation on cigarettes also helps curtail use to some extent. Unfortunately, the economic classes most at risk in our present health care system also tend to be the highest users of tobacco and alcohol.

There is also an ethical dilemma in this area, which is the conflict between individual freedom of choice and harm to society. The same dilemma occurs in providing freedom of choice in prolonging the life of an individual at any cost if he or she can afford it versus providing general health care for the poor.

One of the problems in developing a comprehensive national health care policy is that studies have shown that the non-expert's perception of risk is governed not by numbers and statistics — in other words, by facts — but by subjective evaluation. The adage "Don't confuse me with facts, my mind's made up" has now been proven in a variety of studies. "It won't happen to me, I know people who have smoked for 50 years and are still alive" is a common response of smokers when warned of their much greater risk of heart or lung disease or cancer.

The focus of the media largely controls public perception and is the basis for most people's subjective evaluation of risks. The media can help generate millions of dollars for a specific individual or cause if they can identify an emotional linkage to their viewers. To change our health care system will first require making the need for change newsworthy. This is beginning to happen. Major national news magazines, such as *Time* and *Fortune*, have devoted special issues to the problem, as have national TV networks. However, the media attention waxes and wanes in cycle with public opinion as to the importance of the subject.

Seldom do the media address the enormous costs associated with prolonging the life of the terminally ill for a few more days or weeks. The next chapters, on birth control and death control, will address this issue in some detail.

I Am Owed a Long Life

The second major factor contributing to escalating medical costs is our legal system. People have come to believe that a long life is a right. The US Bill of Rights is no longer "The rights of life, liberty, and the pursuit of happiness". It is increasingly interpreted as "The rights of *long* life, liberty, and the pursuit of happiness".

Rights are protected by law. If your rights are violated you can sue for damages and penalties. Consequently, it is progressively more common for individuals to sue their physician, hospital or health care company when something goes wrong. There is an ample supply of attorneys in the US who are willing and perhaps even eager to prosecute such cases. There is a rapidly increasing trend towards medical litigation in the UK as well. There are four times as many lawsuits now against the UK medical profession as there were just eight years ago. In the US settlements can be in the hundreds of thousands or even millions of dollars. Costs of the litigation including court costs, and legal fees for both parties are usually in the hundreds of thousands of dollars.

This legal surcharge is passed on to the consumer, to each of us. It is passed on as higher physician's fees to cover the explosive growth of malpractice insurance premiums. Charles Breenan reported in the *New York Times* (22 June 1991) that more than $100 billion a year is spent on insurance against malpractice and product liability litigation. It can cost as much to win a case as to lose it. A related article in *Fortune* cited a lower figure, but regardless of the exact cost it is an immense burden on everyone.

This cost is passed on as higher health insurance premiums to individuals and businesses. Health product costs are increased to cover potential legal liability costs. Hospital costs are likewise increased to cover potential legal obligations.

These legal surcharges spread the costs due to the expectation of minimum risk among the entire bill-paying population. The poor and non-insured are protected by the same legal system as the bill-paying population, and their costs as well as their legal surcharge are also added.

In addition to the direct legal surcharge, there is an indirect health care cost incurred to provide protection from litigation. It leads to what is often called "protective medicine". A variety of analytical tests are now available to aid the physician in diagnosis. Each test is costed. Many medical analysis machines are available in every moderate size medical facility. X-ray and magnetic resonance imaging (MRI) machines, CAT scans and ultrasound scans are nearly as commonplace as a stethoscope or a thermometer. However, for the $1,000 cost of one half-hour MRI scan you can buy ten stethoscopes that last a lifetime.

The power of the technology of the analytical machines is twofold. First, there is no question that for many diagnoses the machines supplement the subjective evaluation of the physician and eliminate uncertainty in treatment. However, it is difficult to measure the relative effectiveness of this additional cost to a diagnosis. The cost–benefit ratios are uncertain. Thus, if the physician thinks it will help, he or she may order it regardless of cost.

The second, and perhaps even more powerful, use of the machines is to provide a layer of protection between the physician or hospital and the court.

If a patient does not respond to a treatment and threatens a lawsuit, the presumption of guilt is now shared by the medical team and the machines. Juries are much more inclined to believe that machines are infallible than mortals.

If an X-ray is ordered and no tumour or pneumonia is visible in the lung, the physician is excused for "missing it". If *no* X-ray is ordered and the patient quickly turns worse and dies, the physician may be judged guilty of neglect.

Consequently, risk is minimised for both the patient and the physician if a full work-up of blood tests, urinalysis and X-rays is ordered. If there is any residue of doubt, MRIs are added to the list. In order to ensure the availability of the machines, physicians are encouraged or tempted — depending on your perspective — to invest in diagnostic centres which own and operate the machines. The ethics of a physician profiting from a patient's diagnosis is actively debated. The debate must recognise that legal protection is a major factor in the proliferation of such diagnostic centres as well as profit.

Another less visible but ubiquitous contribution to high health costs is paper work. This component is also indirectly, and in some cases directly, linked to the legal system. The days are long gone when a primary care physician could listen to a few coughs, palpate the chest, look at the fingernails, eyes and ears, examine a chart recording the patient's temperature, pulse and blood pressure, make a diagnosis, write a prescription, and say, "See me in a few days if you're not feeling better," and move on to the next patient.

The filling-in of forms and records occupies a progressively larger fraction of the physician's time and requires adding accounting staff. Chasing insurance claims has become as important as curing patients. Many physicians retire or cut back patient load or eliminate certain types of patients in order to cope with the problem.

This explosive growth of paperwork expands throughout the health care network. Paperwork costs in the US are documented to be more than ten times those in Canada for equivalent services, according to Professor Evelyn Shapero of the University of Manitoba. For example, in the US in 1990 it cost $1.76 to process a Medicare claim. In Saskatchewan the comparable cost was 14 cents. A decade later, the costs of paperwork have increased even further.

Our expectation of a cure at all costs leads to a multiplication of these two factors, i.e. the costs of expanding medical technology are multiplied by increasing legal costs. The cumulative effect is more expensive diagnoses, longer and more heroic efforts to save lives, protection to ensure that all possible risks are minimised, documentation in quadruplicate, and health care insurance that covers an ever-decreasing fraction of the population and their illnesses.

I'll Pay to Have a Long Life

The third contributing factor to rising health care costs is an economic system which perpetuates and expands the expectation of a cure. The reason is simple. There is a lot of money to be made in health care. That is the fundamental reason that health care is the most rapidly growing industry in the US. Large sales and large profits quite naturally lead to the fourth factor, greed, which feeds like a shark on the other three.

A friend of mine in the biomaterials field made $4 million or more bringing a product to the market that he privately admitted to be "garbage". Some patients benefited. Quite a few had questionable results. All had expensive bills. Two faculty colleagues of mine took early retirement from university teaching posts, in part because they could triple their income doing medical liability consulting. Fees of $1,000 per day plus expenses are commonplace for good expert witnesses in such cases. This is a lot more profitable than grading engineering lab reports, and less frustrating.

Two medical products salesmen I have known personally maintained that incomes of $250,000–$500,000 per year or more are commonplace in their field. One of them, who I mentioned earlier went to federal prison for corporate fraud, honed his arts of persuasion and deception as a medical products salesman. It is thoroughly documented in a notorious New York court case, and later in his book, *The Salesman Surgeon*, that he assisted several orthopaedic surgeons in the operating room using a new type of hip implant he had sold the hospital. He claimed that in difficult cases he was a lot better than some of the surgeons, especially when things went wrong, because he didn't have a reputation to worry about. He claimed to have learned his surgical skills by practising on cadavers obtained by bribing hospital attendants and smuggling the dead bodies in the back of his Mercedes home to his garage.

In return for ordering implants from him, he would allow surgeons to practice on the cadavers in his garage late at night. His lucrative and aggressive salesman–surgeon empire was doomed when serious complications developed in a patient. The family's lawsuit exposed hospital records that implicated the salesman and the surgeons. The practice was obviously stopped and in this case the legal system did eventually protect subsequent patients from the greed stimulated by the overall system. The salesman escaped punishment, however, by serving as a government witness and graduated to higher levels of more lucrative crime, corporate theft and stock fraud. His greed and that of a number of associates led to hundreds of investors losing a major part of their retirement funds and deprived thousands of patients of an important new technology that had been funded by government support.

An indication of the amount of money involved in health care is the $5 billion/year annual sales of one medicine, Zantac®, for prevention and control of stomach ulcers and gastritis. A rival, Tagamet®, still has sales of nearly $2 billion. There is no question that the profits from these huge volumes of sales provide the drive for improving medicines and technology. It was the enormous success of Tagamet that stimulated the development of Zantac, which is claimed to be more effective for some patients.

A pharmaceutical success brings about improvement, which stimulates more improvement, and the cycle repeats, benefiting the consumer, the patient. Where is the flaw? Where is the problem?

The trouble with the system is the rising expectation of a "quick fix". If you hurt, take a pill. If it still hurts, take two pills. If your hip or knee hurts, get an implant. If something goes wrong, sue. If you don't win, appeal. Always find someone else to fix it or to blame.

Lasch's book published in 1991, *The True and Only Heaven*, develops this viewpoint. He makes the point that more and more people expect a life of affluence and fulfilment. Charles Breeder's essay in *The Times* summarises, "Because younger Americans are no longer taught deeper values, they expect to be handed immediate gratification. They start searching for somebody to blame when they discover they can't have it all."

Gail Wilensky, director of the US federal agency that runs Medicare and Medicaid, was quoted in *Fortune* magazine as saying, "People want the health care systems to undo all the damage they heap upon themselves."

I submit that this analysis is especially valid for health care and life expectancy. We demand quick cures. Billions of dollars are spent annually on instant diet formulas. However, more than 25% of Americans are overweight and many are clinically obese. Billions are spent on tranquillisers, sleep aids, pep pills and stimulants; all are presumed to provide a quick cure. The difficulty of quenching hard drug usage in the US is in part the dominance of "quick-fix, soft drugs" in our culture. *Fortune* says treating alcohol and drug abuse is the fastest-growing item on corporate medical bills.

There is little attention given by the government, communities or the media to eliminating the need or desire for quick fix solutions to health problems.

If one quick fix doesn't work, there's always a new one coming on the market to try.

About 20 years ago I experienced a series of health problems due to stress which led to several periods of hospitalisation. After finally listening to the advice of several physician friends, I made several adjustments to my life, modified my career goals, and changed my priorities. It was not easy but I was warned the alternative could be fatal. The problem is still there but it is under control and has not required any further trips to the hospital. Several of these friends said 90% of their patients, including me, had gastro-intestinal or nerve-related problems that could be eliminated by altering their behaviour and life-styles. They said nearly all their patients were rushing too much, were trying to do too much at work, were spending too little quiet time alone or with their families. Every day was consumed with meetings or appointments. Nearly every evening was spent with a briefcase, church or social work of some sort. It was always, "more, more, more". The university, business and even churches are equally guilty of such demands.

Busy schedules are often passed on to children, who also are expected to excel in sports, clubs, church and school. When the kids rebel the parents consider themselves to have failed and try even harder. The cumulative effect is a variety of health problems for both the parents and the children. Rather than recognising the source of the problem, quick fixes are requested and usually given.

The sickness of national health care systems is similar. Most discussions focus on a quick fix of parts of the system rather than addressing the fundamental problem of unrealistic expectations, liability if the expectations are not met and cures at any cost.

Human Error, a Magnifying Factor

Each of the issues described above which affect health care costs and distribution leads to human errors. The errors magnify the costs and impede distribution. This compounds the problem. The US Institute of Medicine established a Quality of Health Care in America project in 1998, with the objective of developing a strategy to achieve a threshold improvement in the quality of health care in the US over ten years. The first report on patient safety, *To Err Is Human: Building a Safer Health System*, was published in 2000, and it identifies the magnitude of concern. The number of deaths, prolonged illness and costs associated with human error in health care are staggering. The report also recommends numerous steps that can be implemented to reduce errors and thereby decrease premature death, suffering and cost.

It is estimated that 44,000–98,000 Americans die each year as a result of medical errors. The report concludes, "More people die in a given year as a result of medical errors than from motor vehicle accidents (43,458), breast cancer (42,297) or Aids (16,516)." The total national costs of preventable medical errors resulting in injury or death are estimated to be between $17,000,000,000 and $29,000,000,000 per year, of which half are health care costs.

The report makes an important point that errors are also costly in terms of loss of trust in the health care system by patients and families and diminished satisfaction for both patients and health care professionals. When errors occur there is additional physical and psychological damage to patients that often magnifies their original ailment. Health care professionals lose morale and experience frustration at not being able to provide the best care possible. Blame becomes rampant and prevents the dialogue required to create long-term solutions, locally and nationally. A fresh approach is required where errors are minimised and the emphasis shifts from blame and lawsuits to enhancing the quality and coverage of care for all the people.

Health Care Cure

In this final section let us examine some of the plans developed to achieve universal health care insurance in the US and see how they address the problems of availability of care and escalating costs. The 15 May 1991 issue of *The Journal of the American Medical Association* was devoted to proposals for resolving the problems of the US health care system. Eighty proposals were submitted and after rigorous peer review thirteen were published. The proposals generally followed one of four approaches, as summarised by R.J. Blendon and J.N. Edwards in an editorial in the *JAMA* issue:

(1) A compulsory, employer-based private insurance programme, with the government insuring non-workers and the poor;
(2) A plan that requires employers to provide their employees with health insurance or pay a tax, with the government insuring non-workers and the poor;
(3) A programme of income-related tax credits for individuals, independent of their employers, for the purchase of private insurance;
(4) An all-government insurance system.

There are strong advocates for each of these four approaches. There are also strong arguments against each approach. Consequently, governmental action is delayed until all of the lobby groups have been heard. This process can never reach consensus. The issues are too complex and the stakes are too high. The dilemma has prevented the achievement of a national policy for health care in the US for several decades.

In September 1990, the US Bipartisan Commission on Comprehensive Health Care — the Pepper Commission, chaired by Senator John D. Rockefeller, IV, issued a call for action to address the growing health care crisis. The Commission included twelve members of Congress (six from the House and six from the Senate) and three presidential appointees. Its goal was to develop recommendations for workable and, most importantly, enactable legislation that could solve our health care problems. After extensive hearings and briefings and deliberations for a year, the Commission prepared a blueprint for reform, summarised by Senator Rockefeller in the 15 May 1991 *JAMA* issue on national health care.

The Commission's recommendations rested on four fundamental conclusions:

(1) Health insurance coverage must be universal. Cost shifting and underservice to the uninsured can be avoided only if all Americans are assured of access to care when they need it.

(2) Patching the current system with expanded Medicaid coverage cannot achieve universal coverage. Insurance coverage is simply too expensive for the uninsured and the underinsured using the present system, where employers can elect to provide coverage or not.

(3) It is not practical to replace the current system with a government-run national health insurance. It is not feasible politically to shift millions of people and billions of dollars from the private sector to the public sector.

(4) Controlling costs must go hand in hand with expanding access. The system will likely fail altogether unless the two goals are achieved together.

Thus, the Pepper Commission concluded, "We must secure and extend the combination of job-based and public coverage we now have into a system that truly guarantees adequate coverage for all Americans and that ensures effective and efficient operation in private and public coverage alike."

In order to achieve this goal the Commission recommended enacting special measures to lower the barriers to voluntary purchase of insurance that small businesses face.

Almost all businesses with more than 100 employees provide health insurance from the private sector. Employers with fewer than

100 workers employ about two-thirds of the working uninsured. At present, insurers increasingly compete to insure the young and healthy low risk groups and avoid workers who have had illnesses or are seen as likely to require medical care. The result is denial of insurance to those groups, or termination of coverage or substantial increase of rates for certain employees. Small companies are thereby excluded from employee protection by these practices. As a result, millions of workers who were once well insured face the possibility that coverage will not be available when they need it most.

To counter this trend, the Pepper Commission recommended reforms in the private insurance market that would guarantee the availability of a specified minimum benefit package to all small businesses and their employees. To ease the burden of health insurance costs the Commission also recommended tax credits for small employers.

In order to guarantee the availability of coverage, the Commission's recommendations would give employers a choice: purchase private coverage at whatever rate can be negotiated or purchase coverage from a newly established federal programme. That programme could be administered in conjunction with or as part of Medicare through private insurers or states, under federal rules.

To cover non-workers and thereby guarantee coverage to all, the Commission suggested replacing Medicaid with a new federal programme for non-workers and the self-employed. The same federal programme would be available for employers who found it more affordable. Senator Rockefeller summarised, "The federal program would provide the same minimum benefits that employers must provide. The requirements would ensure national standards for eligibility, benefits, and payment that, in contrast to Medicaid, would guarantee all Americans, no matter what their income, employment status, or place of residence, access to affordable insurance protection."

The minimum standard recommended by the Commission was similar to but less generous than that offered by many employers. It would guarantee access to primary as well as catastrophic care for the uninsured. It would include cost sharing adjusted to ability to pay so as to keep consumers aware of costs, and, importantly, it would emphasise preventive services.

For example, the Commission plan would include hospital care, surgical and other in-patient physician services, diagnostic tests, and limited mental health benefits. Preventive benefits would include prenatal care, well-child care including immunisations, Pap smears, colorectal and prostate screening procedures, and other cost-effective preventive services. Senator Rockefeller concluded, "The emphasis on preventive services reflects the Commission's view that early diagnosis and treatment may result in reduced mortality rates and increased quality of life and may produce savings by averting the cost of expensive treatment."

In order to pursue cost containment the Commission recommended insurance reform that would prohibit competition based on a quest for good risks and would promote competition among insurers to manage care efficiently. The Commission would require that insurers who offer managed care to large businesses extend it to small businesses.

Public controls of costs would be extended. Medicare rules such as prospective payments for hospitals and a resource-based relative value scale for physicians would be used as a model for private payment as well. Also, the federal government would extend data collection, outcomes research, and provide practice guidelines and quality assurance mechanisms to help public and private purchasers to use their money wisely.

Unfortunately, the Pepper Commission was unable to come up with an effective plan to deal with the problem of medical malpractice. It recognised the importance, as indicated by Senator Rockefeller: "Malpractice litigation, as currently handled, fails to protect patients and burdens the health care system with high premiums and the costs of defensive medicine." Since no consensus could be reached, its solution was to refer the issue to more commissions and congressional committees. This is a serious problem, because, as we discussed earlier, without resolution of the legal issues that amplify the costs of care it will be impossible to stop the escalation of costs.

Many of the proposals presented in the special issue of *JAMA* incorporated cost controls. Dr George Lundberg's editorial summarised the controls that could be put in place. They included some severe measures, such as: limitation of the number and types of physicians, a legislated maximum percentage of GDP for health care, overall national or state medical expense caps, and resource-based relative value scale payment for physicians from all payers. Other controls suggested were less controversial, and probably less effective. They included: higher deductibles and co-payments to increase patient restraint in demanding care, pre-use approval for expensive technology, diagnosis-related groups for all hospitals and payers, and better education of physicians and patients about spending options.

There is resistance by many parts of the health care and insurance sectors to any, or perhaps all, reforms. Numerous efforts have been made. So far all have failed. The first major initiative of the Clinton administration in the early 1990s was health care reform. The dismal failure of the legislation was blamed on "Too Much, Too Fast". The reality was the inability of the Republican and Democratic political parties to reach a compromise that would benefit both parties.

Political infighting over health care reform has continued to the present. A headline in *USA Today* (13 April 2000) proclaimed: "Bush, Gore plans fail to hit balance for uninsured." The editorial's lead paragraph said, "At last count, 44 million Americans lacked health insurance, leaving them to suffer through minor illnesses in misery and tolerate major ones until they become critical [Figure 8.3]. At the same time, the majority of those who do have health insurance are locked into plans that are frequently inadequate, inefficient and downright hostile to needed care."

The editorial went on to analyse the health care plans proposed by the leading candidates for the US presidency: Al Gore, Democrat, and George W. Bush, Republican. The conclusion is that "Bush relies mainly on tax credits, which promote choice but are inefficient. Gore

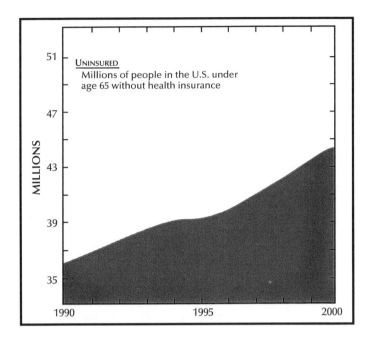

Figure 8.3 Increase in the number of people in the USA who do not have medical insurance.

emphasises expansion of government-run programs, which are more efficient but provide fewer choices."

Both parties would argue vigorously with this conclusion. Republicans maintain that private health care plans are always more efficient than government-run programmes. Democrats proclaim that private programmes always favour the well-to-do and do not protect the poor. The result is an impasse, the same deadlock that was present when the bi-partisan Pepper Commission met a decade ago.

A consequence of this continuing impasse is a growing moral dilemma for physicians and their patients. A recent report in *JAMA* by Matthew Wynia and Ian Wilson concludes that physicians are driven more and more to medical civil disobedience. A random sample of 720 doctors revealed that 40% admitted to deceiving insurers "sometimes" or more often. They willingly exaggerated illnesses, documented false symptoms and changed diagnoses in order to obtain insurance payments for their patients.

"It's an indictment of the system that a physician cannot do right by patients and play fair with insurance companies," says Wynia.

An editorial on this report by Ellen Goodman of the Washington Post Writer's Group says, "Historically, we trained doctors to think of patients one by one. We want them to put us first. Yet as costs rose, we gave insurers — private and public — the social task of keeping the lid on, passing out limited funds and rationing healthcare.

"We never meant to have a system in which you get what your doctor fights for or fudges for. We never meant to have a system based on subversion rather than discussion.

"While nearly 40% of doctors feel they have gamed the system, only 15% agree that it is ethical. Call it a 'game' if you will, but in medicine the moral gray area is not a comfortable playground."

It is time to change the game. It is time to rethink the relationship between physician, insurer and patient.

Time for a Change

The time has come for a change, a big change. Bruce Hilton asks, "Is health care a saleable luxury, like tanning parlours, Caribbean cruises, and cars that will do 120 miles an hour?"

"Or, is health care an essential of a humane community, like roads and public schools — a service that we all band together to pay for because we owe it to one another?"

Hilton goes on to emphasise that health involves the whole person and we must return to a medical approach that recognises the importance of maintaining harmony of mind, body and spirit.

He says, "Humankind has known this for thousands of years. It is central to most religions. The idea permeates the Hebrew and Christian scriptures and is central to the Christian faith. It's no coincidence that whole, holy, and healing have the same word root.

"So, it's more complicated than it looks at first. Wrestling with health issues means more than struggling with the dilemmas raised by the technology of modern medicine, as useful and God-given as that medicine may be.

"It also means helping to create loving, trusting communities — in our families, neighbourhoods, nations, and world."

Daniel Callahan also concludes in his book that we must return to a system where the emphasis is on care of the person rather than cure of an organ.

We come to the end of this chapter with a growing recognition that a major readjustment of attitudes towards health care is needed.

We need to reduce our expectation that ills can be cured with a quick fix.

We must accept that terminal illnesses should not be prolonged unnecessarily.

We must accept uncertainty in life and recognise that all medical treatment involves risk.

We need to return to the attitudes of the past, where care was as important as cure.

Technology must be considered to be only a component to restore health rather than an alternative for maintaining health.

We must re-establish our values and recognise that good health requires good relationships within the family and between ourselves and our fellow workers.

We must also remember that life is a gift from God and must reaffirm that our relationship with our Creator is the foundation for a good life and good health.

Likewise, we must accept that life is limited and when it is time to go technology is no substitute for God and God's will.

Bibliography

1. *The BMA Guide to Living with Risk*, The British Medical Association, Penguin Books, 1990
2. *What Kind of Life: The Limits of Medical Progress*, Daniel Callahan, Simon and Schuster, 1990

3. *Journal of the American Medical Association*, 15 May 1991

4. *Fortune*, 1 July 1991

5. *First Do No Harm: Wrestling with the New Medicine's Life and Death Dilemmas*, Bruce Hilton, Abingdon Press, 1991

6. *Business Ethics, Corporate Values and Society*, Milton Snoeyenbos, Robert Almeder and James Humber (eds.), Prometheus Books, New York, 1983

7. *Ethics, Computing, and Medicine: Informatics and the Transformation of Health Care*, Kenneth W. Goodman (ed.), Cambridge University Press, Cambridge, 1998

8. *Long-Term Care in an Aging Society*, Gerald A. Larue and Rich Bayly, Prometheus Books, New York, 1992

9. *The Rise and Fall of Modern Medicine*, James Le Fanu, Little, Brown and Co., London, 1999

10. *Risk: Analysis, Perception and Management*, The Royal Society, London, 1992

11. *Science, Policy and Risk*, The Royal Society, London, 1997

12. *To Err Is Human: Building a Safer Health System*, Linda T. Kohn, Janet M. Corrigan and Molla S. Donaldson (eds.), National Academy Press, Washington D.C., 2000

9
Ethical Issues of Birth Control

The two most emotional times in our existence are at the beginning and the end of a life. We hold a newborn infant in our arms and wonder at the beauty and mystery of the new creation. A new baby is both a wonder to hold and a wonder to behold. The little fingers down to their tiny nails, the eyes and eyelashes, the unblemished skin — all perfect replicas of the mother and father, or at least a nearly perfect, nearly replicated version of its parents. Differences are always present. Sometimes the differences are visible, such as red hair and green eyes in my case. Sometimes the differences do not appear for many months or even years.

Sometimes the differences are good. The infant may develop a unique ability in athletics, music or mathematics far beyond that of either parent. Sometimes, the differences are not so good. The infant may have physical problems such as crossed eyes, a heart murmur, problems with its lungs (hyaline membrane disease), spina bifida (an exposed spinal cord) or paraplegia. It may have mental problems which only show up later.

All of us, either as parents, relatives or friends, rejoice when the differences are good and despair when they are bad. We give thanks to the Lord for answering our prayers when they are good, or we cry out in a lamentation like Job when they are bad: "Why, Lord? Why produce this suffering? What is the meaning? What did this little one or its parents do to deserve this tragedy and suffering?" We seldom receive an answer and must live with the reality and accept the uncertainty.

Just as the origins of life some three billion years ago are uncertain and shrouded in mystery, so too the beginnings of a new life today are filled with mystery and uncertainty. At least that has been the case for millennia upon millennia, for species upon species.

Uncertainty and differences have been at the very heart of the natural order of life. As we discussed in Chapters 3 and 4, one of the three requirements for self-sustaining life is mutability, the ability to adapt. The other two requirements for life, self-maintenance and self-replication, are sufficient as long as the external world remains relatively constant. However, as we experience continually, the physical world is never constant, never has been and — by extrapolation, the basis of science — presumably never will be. Volcanoes erupt, hurricanes and cyclones roar, rivers flood, land freezes, rain fails to come, meteorites crash into the earth, oceans rise, fall and even disappear. Even our sun will one day grow cold and collapse, as have countless other stars. The occurrence of all these physical events is unpredictable and uncertain. The only certainty is that change is inevitable. Life must be able to respond to change when it occurs or it will die out. That is the way it always has been and always will be.

Or is it?

Humans now have the capacity to alter this natural order. For quite a long time people have been able to moderate to some extent the changes that occur in the physical world. Rivers are dammed and lands are irrigated, buildings are constructed to provide shelter in the severest of storms, crops and animals are raised and harvested in great quantity and with great predictability. No longer is the availability of food, shelter or clothing subject to the vagaries of nature for most people, especially those in the Northern Hemisphere. When the necessities of life are missing, such as is still the case in much of Africa, the problems are distribution and political as much as natural. Thus, technology is responsible for eliminating or minimising the uncertainty in obtaining the necessities of life for most of us.

The influence of technology on the natural order has also had a profound effect on the uncertainties associated with both the beginning and the end of life. Humankind has experienced three revolutions in the control of birth, all three based upon technology.

The first revolution was the recognition of the importance of bacteria and the ability to achieve clean conditions and the use of disinfectants and antibiotics to combat infection of either baby or mother. The result has been that now ten times as many babies survive and few mothers die in childbirth. Thus, the first revolution in birth control was a revolution of maintaining life at birth, or *life preservation*.

The second revolution has occurred within our lifetimes. It is a revolution of *life prevention*. Although contraception has been practised throughout the history of humans, it was always limited and the results were highly unpredictable. The prevention of new life was never the norm. It simply was neither possible nor desirable, since the rate of death at birth was so high. Large families were needed to hunt or to work the farms and maintain the family.

In many ways the second revolution is an outgrowth or consequence of the first. The earth cannot withstand an exponential growth of population, which is the natural consequence when the natural barriers to early death are largely eliminated (Figure 9.1). Thus, today, the social norm for most of the six billion people in the world is birth prevention in one form or another. Even when the official position of a church is opposed to the practice on doctrinal grounds, most communicants still practice birth prevention based on moral and practical grounds.

Most people believe that the moral weight is on the begetting and rearing of children that are wanted and can be cared for rather than accepting a "natural" or God-given number. Few churches are opposed to birth prevention if it is achieved by natural methods, such as abstinence or use of the infertile period.

During the last 40 years technology has produced various means for overriding the natural order of conception. These methods are generally safe and predictable, at least 90% or more of the time. Although the long term effects of the Pill on society are still debated, as discussed in many of the volumes listed in the bibliography, the

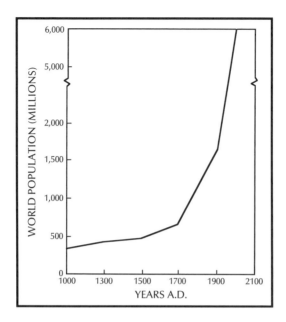

Figure 9.1 Increase in the world population during the last millennium.

reality is that it has been the major factor in the revolution of birth prevention. The Pill has given women control over when they will conceive.

Now, before society has adjusted to the second revolution of birth prevention, biotechnology has thrust a third revolution upon us. It is more disturbing. It is a revolution that has the potential of complete birth control. Control not just when women *may* conceive but of *what* they may conceive.

It is a revolution of *birth intervention*.

There are almost no moral and ethical issues associated with the first revolution of birth preservation. Nearly everyone will agree that the saving of the life of a newborn infant and/or its mother is good, in the most fundamental way.

Likewise, there are relatively few moral and ethical disagreements generated by the second revolution, birth prevention. The moral principles of respect for autonomy, beneficence and justice are generally preserved, and in fact often enhanced, when a man and a woman purposely decide together and in harmony not to conceive a child. The teachings of the Bible, Torah or Koran, the Ten Commandments, and Christ's Great Commandment, are all consistent with a decision to restrict a family to a size that can be provided for with loving care.

However, the third revolution, birth intervention, has become a major divisive force in society. It makes Pandora's box seem like a can of worms. Pro-right-to-life vs. pro-right-to-choice has been the most hotly contested moral issue in the US since slavery. Taking a stand on abortion is the surest way to end a political career or ruin a dinner party.

Approval of the RU 486 "morning-after pill", which ends the need for surgery in early pregnancy, was granted by the British government in 1991. The news, reported by Caroline Lees in *The Times*, was greeted with enthusiasm by British doctors and family planning clinics. At the same time anti-abortionists met to plan a national boycott of the manufacturer, Roussell, which also makes veterinary products, insecticides and gardening equipment. The national campaign co-ordinator of the anti-abortion group said it was planning its biggest campaign ever against the drug. "We will try everything to stop this pill," he said. "We see it as a return to the days of back street abortions, except that dirty knitting needles and caustic soda will be replaced by dangerous chemicals. We are horrified that pharmaceutical companies are making drugs that are designed to destroy lives, not save them."

The French biochemist Professor Etienne Eonilie Bauilieu, who invented RU 486, defended it saying, "My mission is to help women. There is great need for this pill in developing countries and in Eastern Europe, where the situation is tragic; there is little contraception and abortion is still often illegal."

The morning-after pill was introduced in France in the late 1980s, where it is used for about one-third of early abortions. About 2,000 British women used it in controlled trials, with 94–96% success. It makes outpatient treatment possible for women wanting to terminate a pregnancy before the tenth week. Women require three hospital visits, including a check-up, but the pill eliminates the need for surgical procedures, general anaesthetics and overnight stays in hospitals and clinics. The pill is administered and supervised by the hospitals and clinics licensed under Britain's 1967 Abortion Act. Surgical intervention will still be available up to 28 weeks. Over-the-counter approval was granted in the UK effective January 2001.

The impact of the pill on health care cost in Britain should be substantial, since more than three million terminations have been carried out in England and Wales in the past 20 years. The numbers have increased dramatically. There has often been a one-month waiting list for surgical abortion. That is bad for the woman. In 1989 185,000 women had terminations compared with 54,000 20 years earlier.

Due to technology the level of safety is now sufficiently high that the risk associated with intervention to terminate an unwanted pregnancy is judged by thousands of women to be more acceptable than the financial and emotional consequences of caring for the child. Of course the child-to-be has no say in the decision, which is central to the moral and ethical dilemmas associated with this third birth control revolution, birth intervention.

The emotional fervour of abortion runs high. However, it is essential to recognise that it is only one of the slippery worms in birth intervention. Many others are equally disturbing in their social and moral consequences. Consider, for example, some of the issues associated with *in vitro* fertilisation (IVF) and artificial insemination by donor. These technological advances were stimulated by the seemingly morally justified desire of a childless couple to have and rear a child.

Where's the problem? One concern is the potential to use genetic engineering to change genes in humans that cause various genetic defects. For example, it is conceivable that an embryo could be altered to eliminate the defective gene that is responsible for diabetes, a disease that affects some four million clinical diabetics in the US. The gene therapy would free them from a lifetime of insulin injections. Where's the problem?

The problem is one the moral philosophers call "a slippery slope problem". What that means is, once you take one step down that slope, you cannot predict how far you will slide. Many of the essays in *Ethics, Reproduction and Genetic Control* (edited by Ruth Chadwick), or Chapters 4, 5 and 10 of *Contemporary Issues in Bioethics,* 3rd edition (edited by Tom L. Beauchamp and LeRoy Walters), discuss these slippery slope issues.

Robert L. Sinsheimer, in the Chadwick volume, introduces the slipperiest of the worms, the potential for genetic manipulation of humans, in a highly poetic manner: "It has long been apparent that you or I do not enter the world as unformed clay compliant to any mould; rather we have in our beginnings some bent of mind, some shade of character. The origin of this structure — of the fibre of this clay — was for centuries mysterious. In earlier times men sought its trace in the conjunction of the stars or perhaps in the momentary combination of the elements at nativity. Today, instead, we know to look within. We seek not in the stars but in our genes for the herald of our fate.

"Today there is much talk about the possibility of human genetic modification — of desired genetic change, specifically of mankind. This concept makes a turning point in the whole evolution of life. For the first time in all time a living creature understands its origin and can undertake to design its future."

Talk about a slippery slope. There are few steeper. Think back to Chapters 3 and 4, where we discussed the origin and evolution of life. "Genesis" 1, 2 says, "And, God created humankind in His own image. In His own image did he create man and woman. And, God saw that it was good."

Where's the problem? you ask. The problem is that the slippery slope begins with God the Creator and man the created and slides quickly and ominously to God the Creator and man the manipulator.

I ask, "If we presently do not understand and comprehend the origin of life, how far then can we justify altering the product of that origin? Does not the uncertainty inherent in our beginning tell us that there must also be uncertainty in our ending? How much of an attempt to eliminate that uncertainty can be accepted before the slope gets too slippery?"

For many people the most immediate reaction to these questions tends to be one of fear. The fear of the unknown. For thousands of years such fear dominated when most humans considered exploring unknown lands. Most people never travelled beyond their own villages. Now, nearly the entire earth has been explored and the problem is exploitation rather than exploration.

Likewise, the fear of nature is now largely the stuff of stories. People hardly bother to look at eclipses anymore, if they interfere with bedtime or a ball game, let alone fear that the moon or sun is being devoured by a dragon. In fact, life is so free of fear for most that the entertainment industry makes billions of dollars per year providing fake fear in the form of amusement rides, movies, TV and books that emphasise terror.

As always, this generalisation is valid only for the majority. If you live in an inner city, or your car breaks down late at night in the Bronx, or you are put in jail, as depicted so graphically by Tom Wolfe in *The Bonfire of the Vanities*, you would experience plenty of real fear first-hand. Or, if you invest your life's savings in stock futures, you experience real fear when the markets fluctuate wildly.

There is no question that the fear of failure is as real now as in the past. The high number of suicides in all developed countries reflects that fact.

However, the point is that the fear of the unknown facing all of us confronted with the revolution of birth intervention is not an individual fear. It is a fear that our entire species may be altered. It is a fear that the natural order may be changed irreversibly.

Many of us have been raised knowing the story of the Garden of Eden and God's warning of the consequences of eating the fruit from the tree of knowledge. We know that it is the destiny of humans to give in to that temptation. We have all seen the consequences of the first two revolutions, birth preservation and birth prevention, and judged them to be good. We know that we cannot resist the challenge of the third, birth intervention. We know that the results are often good. The joy of a childless couple giving birth to an IVF baby is one of the triumphs of modern medicine, as so delightfully described by Professor Lord Robert Winston in his books.

However, we all have to accept the responsibility and understand the potential negative as well as positive consequences. We have to make an informed and intelligent analysis of the issues. We each have to judge how steep the ethical slope is and be prepared to defend how far down to put the barrier and how high to make it.

We have to recognise that uncertainty exists: that it is impossible to predict the steepness of the slope. We cannot, we must not, relegate the decisions on these life-and-death issues to doctors, lawyers and judges. We must become informed and involved as individuals and we must ensure that our churches and synagogues and mosques become involved in the debate and decisions. We must avoid letting emotion hide the facts. We must speak out.

As an example of the complications of birth intervention, let us return to the subject of IVF. As summarised by Sir David Napley, in the Chadwick volume, IVF is concerned with fertilisation in glass, i.e. an embryo produced in a test-tube rather than in a woman's uterus. However, four alternative procedures may be used:

(1) Extract one or more eggs from the woman by surgical or ultrasound methods, followed by impregnation in the laboratory of an egg with her lawful husband's sperm and then return of the fertile egg to the mother;

(2) Same procedure, except for the use of sperm from a donor other than the husband;

(3) Extract an egg or eggs from a third party, a woman donor, followed by fertilisation with the husband's sperm and insertion into his wife's body to eventually produce a child;

(4) Third parties provide both egg and sperm, which are then inserted into the surrogate mother's womb.

Although the technology is the same for all four alternatives, the consequences to the child that is to be born are remarkably different. In the first case, the genetic make-up of the baby will be the same as if fertilisation had occurred normally. There is no reason to expect that the *in vitro* child would be any different from a natural child. This is indeed only a baby step down the slippery slope. If a couple or a society can afford such a process, there is no reason to justify denying a childless couple the use of it on moral or ethical grounds.

The slide down the slope begins with the second and third techniques. The embryo conceived by these *in vitro* methods has only half of the genetic material of the parents who will be rearing the child. In the second method, the mother will have provided half of the chromosomes, but the other half will not be from the husband. These will be from the unknown, anonymous donor. Thus, the child will not know its genetic father.

The third IVF method will result in a child that will not know its genetic mother. The psychological consequences to the child and the child-rearing parents of unknown genetic mothers and fathers are unpredictable. Historically, children born with these genetic conditions have been called "born out of wedlock". The children have been labelled bastards or illegitimate children of adulterers. Often such children have been deprived of certain legal property rights as a consequence of their unknown or "non-lawful" genetic origin.

The fourth method, implanting an egg from a donor woman fertilised by a donor male's sperm, results in a child with no genetic heritage from its legal parents. The last three methods create the slippery slope question: "Is it morally right or wrong to decide to bring a child into being who will not know its genetic origin?"

Ruth Chadwick says, "There will be a difference of view here between those who think that the genetic origins are fundamentally important and those who believe that what is more important is the loving nature of the child in a stable marital relationship." Individuals who emphasise the importance of genetics, such as Robert Graham, who founded the sperm bank in California from Nobel laureate donors, maintain that we have a moral obligation to improve our genetic stock when we have the opportunity or desire to do so.

The slope becomes steep when concepts of birth intervention are extended in this way. Purposeful attempts to modify the genetics of children, called eugenics, are morally slippery indeed. Replacing a defective gene in the embryo sounds like an obvious benefit to all concerned — child, parents, donors and society. But who is responsible if the intervention is faulty? Is it morally justified to fertilise and incubate several eggs, then test, modify and destroy them until an embryo is produced according to the wishes of the parents-to-be? What are the rights, if any, of the IVF embryos not implanted? Who do they belong to? Man? Woman? Doctor? Society?

If a man and his wife in a stable marriage have the right to receive IVF, does this right extent to an emotionally stable, financially secure, childless single person? Is the right sexless, i.e. does it extend to male as well as female? Does it extend to homosexual couples? Does it extend to cloning individuals, as discussed in the previous chapter? The courts are still debating many of these questions.

The slope becomes ever so slippery, doesn't it? The uncertainties discussed in Chapters 1–4 with the theories of the big bang, the primordial soup and evolution pale into relative insignificance when compared with these issues of genetic manipulation and birth intervention. The danger to society occurs when people acquire a mistaken belief that science has all the answers and provides certainty and predictability in an uncertain world. When that happens controls are lifted and the slippery slide begins. We must never forget that the human roles of science and technology compared with faith and ethics are greatly different.

Science provides an understanding of nature, and technology uses that knowledge. Moral and ethical judgements are required to establish whether the technology is a bridge to a better life, like birth preservation and prevention, or whether it is a slippery slope with unforeseen consequences at the bottom, like birth intervention.

Faith in a Divine Creator who is the source of good; faith in the Revealed Word and its unchanging relevance in a changing world; faith that God cares; faith that there is a meaning beyond the material — faith is all of these and more.

Faith does not necessarily provide all the answers but it does provide the foundation for accepting that answers are sometimes not possible.

Faith also provides the moral basis for determining where to put the barriers on the slippery slope created by technology.

When technology is inconsistent with faith, it is time to draw the line.

When the results of technology are incompatible with the dignity of being made in the image of God, it is time to call a halt.

It is the responsibility of each person to make these moral judgements.

Bibliography

1. *Abortion: Law, Choice and Morality*, Daniel Callahan, Simon and Schuster
2. *Ethics, Reproduction and Genetic Control*, Ruth F. Chadwick (ed.), Routledge, London, 1987
3. *The Ethics of Abortion: Pro-Life vs. Pro-Choice* (Revised Edition), Robert M. Baird and Stuart E. Rosenbaum, Prometheus Books, New York, 1993
4. *Contemporary Issues in Bioethics* (3rd Edition), Tom L. Beauchamp and LeRoy Walters, Wadsworth Publishing Co., California, 1989
5. *Morality and the Law*, Robert M. Baird and Stuart E. Rosenbaum (eds.), Prometheus Books, New York, 1988
6. *Contraception: A Guide to Birth Control Methods*, Vera Bullough and Bonnie Bullough, Prometheus Books, New York, 1990
7. *Ethics in Medicine*, Milton D. Heifetz, Prometheus Books, New York, 1996
8. *Genetic Manipulation*, Robert Winston, Weidenfield & Nicholson, London, 1997
9. *Making Babies: A Personal View of IVF*, Robert Winston, BBC Consumer Publishing, London, 1996
10. *Infertility: A Sympathetic Approach*, Robert Winston, Vermillion, London, 1996

10
Ethical Issues of Death Control

We finished the chapter on the ethical issues of birth intervention with the statement "When the results of technology are incompatible with the dignity of being made in the image of God, it is time to call a halt."

That is what happens today in many cases when life ends. Technology in the form of death intervention has also become a revolution during the last few decades. However, unlike the abortion versus right-to-life issues, the right-to-death debate is quiet and submerged in social and personal taboos. Bringing up the subject of death at a dinner party is even worse than bringing up abortion.

A person's desire for death with dignity most often cannot even be expressed.

Like a foetus, the dying often cannot speak. Their wishes cannot be heard, their pain and suffering cannot be felt.

In many cases friends and family are long gone or long uncaring. Or, even in the best circumstances a loving, caring family may be bound by the medico-legal system to do nothing, to stand by and wait.

As is so graphically portrayed in Raffin, Shurkin and Sinkler's book, *Intensive Care*, once a patient is in an intensive care unit on a life support system there are many legal and moral restraints from removing that support.

The moral barrier on the slippery slope is often placed to protect the system rather than the dying patient.

Defining Death

Part of the problem is the difficulty in defining life and death. We saw in Chapter 3 that there are three requirements for life:

metabolism, replication and mutability. When cells are functioning and dividing, as needed for growth or to replace themselves, an organism is considered to be alive. When cells stop functioning and stop replicating themselves, the organism is dead. For single-celled, simple organisms the time of death and the condition of death are relatively easy to establish. Simple observation is usually sufficient.

For large, complex multi-celled organisms like us, it is much more difficult to be precise regarding the status of either life or death. Quantifying the presence of life or death is even more difficult. What is the problem?

The problem is that life requires self-sufficiency. For uni-celled organisms, like bacteria, there is not much doubt. The cells are usually either alive or dead. However, even single cells can exist for a period of time as sick cells, a condition where they do not function well enough to replicate but still have a partially functioning metabolism.

For humans the cellular conditions for life and death are exceedingly more complex. Each of us is made up of many trillions of cells of dozens of different types, each with a very highly specific function. When we are alive most of these cells are alive and functioning normally. Heart cells are expanding and contracting about once every second. Blood cells are circulating, carrying oxygen to other cells or carbon dioxide away from them. Nerve cells are carrying electrical impulses to the brain, where the stimuli are organised and used to regulate in one form or another the action of most of the other cells.

This is an enormously complex interdependence of cells upon each other. However, repair and redundancy are built in. When you cut a finger you don't die. It takes a major breakdown of many parts of the system, the living body, before all functions are brought to a stop. This complex interdependence or synergism of many specific cell types is why defining life and death for people is so difficult.

In thinking about this problem it is important to remember that the several trillion cells that are "you" were originally just one — the fertilised egg or ovum of your mother. Immediately after fertilisation cell division began, with each divided cell carrying a replica of the chromosomes and DNA of the fertilised parent cell. It takes an enormous number of cell divisions before a self-sufficient organism, a foetus, results.

Part of the problem of defining when life begins is a problem of determining how self-sufficient a foetus must be to survive outside the mother's womb. Technology has created this problem. For hundreds of thousands of years there was no uncertainty. Life began when a newborn baby took its first breath and cried.

No breath — no life, death. Simple. No problem. Tragic to the mother and father, but simple to society.

The same simplicity existed at the end of life. No breath — no life, death.

Today there can be a problem, at both the beginning and the end of life. It may only be a problem for a small percentage of births and deaths, but the associated expense affects everyone. The expense associated with death intervention is responsible for a large part of the escalating costs of health care discussed in Chapter 8.

Robert M. Veatch, in an essay, "Definitions of Life and Death: Should There Be Consistency?", in Beauchamp and Walters' book, formulates the physiological source of the definition problem clearly and concisely. He states, "The interesting moral, legal, and political question concerns what features of human life imply standing so that the claims of an individual are comparable, at least for most purposes, to those of other living human beings," He goes on to conclude, "We know that moral shifts take place when life begins or ends, but it is in principle impossible to determine biologically when those points are. If there is some essential feature that gives individuals moral or legal claims comparable to other members of the human community, loss of that feature implies that one no longer relates to the community in the same way."

Veatch suggests that there are four possible answers as to what that essential feature may be. It is the uncertainty about which of these four alternatives to use to define death that causes the moral dilemma.

(1) A critical feature that is widely accepted to give a foetus moral standing is the development of a fixed genetic code. Veatch says, "Full human standing is seen as coming when the genetic code first becomes unique."

Earlier this uniqueness was thought to come into existence at the time of conception, as soon as fertilisation and cell division had occurred. Now, the work of geneticists such as Jerome Lejuene has led many to argue that a unique genetic code is not present until twinning can no longer occur. This is about two weeks after fertilisation and division.

This conclusion could have a major influence on the moral attitude towards abortion within the first two weeks. Use of the morning-after pill, for example, does not abort unique genetic material.

Fixation of a unique genetic code, however, has little bearing on an appropriate definition of death. A person can show no vital signs of life and still contain cells viable enough for cloning, bizarre as that may seem. Therefore, the feature of a unique genetic code does not give consistency in providing the beginning and end point of human standing.

(2) For millennia the traditional pronouncement of life and death has been the presence or absence of a beating heart and breathing lungs. Technology has destroyed that definition. Life can be maintained almost indefinitely on respirators, on heart–lung machines and heart assist devices. The patient living with the assistance of these devices obviously is not well, but he or she obviously is not dead either.

So, the absence of normal heart and/or lung activity is no longer a definitive feature for establishing death.

There is another concern in using heart and lung function as the essential feature for establishing the presence of life. The problem is the variation in time required to produce a functional, self-sustaining circulatory system. Cardiac tissue begins to be formed in the embryo at around four weeks but does not develop into a functioning circulatory system until the twelfth week or further. Capacity for lung function does not appear in the foetus until six months or more.

Veatch's summary of this position is perceptive: "They, the heart–lung people, have adopted a view that the essence of being a human has to do with something as crass, as animalistic, and as biological as whether liquids are being pumped through the plumbing and gases are properly maintaining the ventilating system. There is nothing unscientific about such a view; it is fundamentally not a scientific judgement. It represents, however, a very limited view of human nature. It rejects the mainstream of the Western Judeo-Christian tradition that says that the human is far more than a mere body. It is even more decisively at odds with the Greek emphasis on the Soul as man's essence, temporarily trapped in the flesh."

Emphasising circulation and breathing as the essence of life is at odds with most Eastern religions as well. It particularly makes a mockery of the teaching of the Bible that "Humankind is made in God's image." The concept of God as a gigantic supernatural heart and lung machine with gasping bellows and wildly pumping blood may be reminiscent of a Woody Allen movie but should hardly be the basis for making life-and-death decisions.

We believe that God is love. We feel instinctively that the essence of being a human must include the capacity for giving and receiving love.

Blood and air may be vital to life but are not alone enough for life.

(3) The third alternative is to recognise brain function as the critical test of life or death. In this view, as stated by Veatch, "The essence of human existence is not related to mere fluid flow but to something far more complicated and subtle having to do with the integration of bodily functions with the nervous system. When that is present we have a 'person as a whole' and not just a collection of cells, tissues, and organs carrying on independent functions. The President's Commission for the Study of Ethical Problems in Medicine and Biomedical and Behaviour Research took this position in July of 1981 when it adopted a brain-oriented definition of death. It has also been officially adopted by 27 states."

The development of a functioning integrated nervous system is likewise a reasonable criterion for life. Nerve cells begin to appear in a foetus at about the third week. However, integration of the nerve cells with organs requires approximately ten weeks. Six months of growth is necessary for nervous tissue to develop interactions with heart, lungs and other organs sufficient for the foetus to survive independent of its mother. This is a similar time period for ascribing the beginnings of self-sustaining life if one takes a heart–lung position.

(4) The fourth position to take in defining the essence of life is to require that the individual not only possess brain function to control organs but also exhibit the capacity for consciousness, for thinking, for feeling, for remembering, for reasoning, for love.

This position, requiring higher brain or cerebral function, is most consistent with the thinking of many of the great philosophers. Descartes said it most clearly: "*Cogito ergo sum.*" ("I think, therefore I am.")

The moral, ethical and legal dilemma is that a person can be alive without thinking. But, can a person be a person without thinking?

This dilemma is not easily resolved. Referring again to Veatch, "We have had considerable uncertainty over measuring the irreversible loss of the capacity for consciousness. Many who in principle favour a concept of death related to loss of consciousness fall back in practice to using the destruction of the entire brain as the clinical point at which we are first really sure that consciousness is no longer possible.... .

"It now appears that there are several plausible positions in the definition-of-death debate that reasonable people may take. I am increasingly convinced that there will never be agreement on any one of these. Some tolerable compromise will have to result, possibly including limited freedom of choice.... . We almost certainly will be forced to live with the frustrating and perpetual uncertainty this creates, just as we live with uncertainty over when we should treat people as dead."

Assisting Death

The moral and ethical dilemma in defining death has grave consequences in treating the terminally ill. A critical issue of continuing debate is physician-assisted suicide. As emphasised in Michael Uhlmann's authoritative volume *Last Rights?*, the moral and religious imperative throughout the Western world has been to condemn assisted suicide.

Uhlmann summarises in his introduction to the third part of the book *Medical Perspectives*, "For two millennia or more, medical ethics has been guided by the moral principles embodied in the Hippocratic Oath. Although it originated in ancient Greece, the Hippocratic tradition was decisively shaped by Judaeo-Christian thought and has enjoyed dominion in Western thought and culture ever since. Under its aegis, medicine was directed toward the good of the patient: first, to do no harm, and where possible, to cure his afflictions. Recognising the unequal power of physicians and patients, the Hippocratic Oath also contained a number of specific prohibitions, including to refuse, even upon request, to take a patient's life.

"The physical and moral assumptions of the Hippocratic tradition are today being challenged as never before. Physician–patient relationships now occur in the shadow cast by moral scepticism, the economics of health care and revolutionary changes in medical technology. These developments create a host of ethical dilemmas that earlier eras never had to confront. Some argue that the moral foundations of the Hippocratic tradition are inadequate to an age that places almost supreme value on personal autonomy."

In other, simplistic words: If you want it, you should have it — including help in ending your life.

Physician Jack Kevorkian espouses this view, both in words and in practice, having "assisted" in the deaths of more than 50 persons. Uhlmann summarises, "In his [Kevorkian's] view, prohibition of physician-assisted suicide is a punitive view foisted upon society by religious zealots who want to commit medicine to, as he puts it, the preservation of life 'at all costs'. In the name of protecting life,

Kevorkian says, the taboo tyrannises patients and exacerbates their suffering."

Kevorkian is not alone in proclaiming the view that when certain conditions are met, people should be free to end their own lives and physicians should be there to help them. Derek Humphrey (Founder of the Hemlock Society), Timothy E. Quill and H. Tristam Englehardt, Jr. all argue in essays in the Uhlmann volume that current restrictions on physician-assisted death operate to the physical and moral disadvantage of patients suffering from incurable illnesses and intolerable suffering. The Dutch agree with this conclusion. The Dutch Government has lifted the ban on physician-assisted euthanasia. Tight controls are in place but the practice is no longer illegal.

Equally compelling arguments in defence of the present barriers to doctors helping patients to die are given by D. Alan Sherman, Herbert Hendin and Edmund D. Pellegrino. For example, Pellegrino reminds us that the debate within medicine mirrors the debate within society at large about the nature of the moral truth. As we saw earlier, moral philosophers cannot agree on the issue of whether moral positions, such as euthanasia, should be grounded upon culture and tradition or whether there are truths that transcend our particular time and place.

Leon R. Kass's essay in *Last Rights?* takes a strong position that there is no moral dilemma. He maintains, from the perspective of both a physician and a philosopher, that "The taboo against doctors killing patients (even on request) is the very embodiment of reason and wisdom." He states, "Even the most humane and conscientious physician psychologically needs protection against himself and his weaknesses, if he is to care fully for those who entrust themselves to him."

Kass proceeds to describe the ethical basis for the barrier against euthanasia, "The deepest ethical principle restraining the physician's power is not the autonomy or freedom of the patient; neither is it his own compassion or good intentions. Rather, it is the dignity and mysterious power of human life itself and therefore also what the

Hippocratic Oath calls the purity and holiness of the life and art to which he has sworn devotion."

This likewise continues to be the position of the American Medical Association summarised by their 1996 President, Lonnie R. Bristow, testifying before the United States Congress (pp. 399–406 of *Last Rights?*).

These positions against physician-assisted suicide essentially say that technology does not make something right that has been wrong for 2,500 years. People have always suffered before dying. The fact that there are many more people today to suffer and to die does not make it morally acceptable to change the sanctity of the doctor–patient relationship.

The decision as to when to die should best be left in the hands of God, not in the hands of man.

However, our concerns about death control should not be limited to the time of death. They need to include the quality of life during our final days leading to death. An outcome of the debate on the ethical issues of death control should be a shift in our attitude towards death and dying. We must accept, honestly and openly, that dying is part of life. Protracting life at all costs should become accepted as morally wrong. A report from the Institute of Medicine (IOM) in the US, "Approaching Death: Improving Care at the End of Life", takes an important step towards bringing discussion of the medical, emotional, legal and economic issues about the ending of life into public debate. The 1997 report concludes, "In principle, humane care for those approaching death is a social obligation as well as a personal offering from those directly involved. In reality, both society and individuals often fall short of what is reasonably — if not simply — achievable. As a result, people have come both to fear a technologically over-treated and protracted death and to dread the prospect of abandonment and untreated physical and emotional stress."

The IOM report goes on the emphasise several themes that are essential if we are to improve the quality of life at the end of life. They are:

(1) "Too many dying people suffer from pain and other distress that clinicians could prevent or relieve with existing knowledge and therapies.

(2) "Significant organizational, economic, legal and educational impediments to good care can be identified and, in varying degrees, remedied.

(3) "Important gaps in scientific knowledge about the end of life need serious attention from biomedical, social science, and health services researchers.

(4) "Strengthening accountability for the quality of care at the end of life will require better data and tools for evaluating the outcomes important to patients and families."

Even if these goals are met, it is during these final days of life that science is often a poor substitute for faith. For those of us that have had religious experiences that transcend the physical world, we can anticipate that these experiences are perhaps a window on the world to come. The religious teachings that we believe to be the revealed Word also reinforce a belief that there is meaning beyond the material. The uncertainty that we face as life comes to an end can thereby become displaced towards acceptance. I believe that science cannot provide that comfort at the end.

Bibliography

1. *Euthanasia: The Moral Issues*, Robert M. Baird and Stuart E. Rosenbaum (eds.), Prometheus Books, New York, 1989
2. *Intensive Care: Facing the Critical Choices*, Thomas A. Raffin, Joe N. Shurkin and Wharton Sinkler III, W. H. Freeman and Co., New York, 1989
3. *Contemporary Issues in Bioethics* (3rd edition), Tom L. Beauchamp and LeRoy Walters (eds.), Wadsworth Publishing Co., California, 1989
4. *Morality and the Law*, Robert M. Baird and Stuart E. Rosenbaum (eds.), Prometheus Books, New York, 1988
5. *Ethics in Medicine*, Milton D. Heifetz, Prometheus Books, New York, 1996
6. *Last Rights? Assisted Suicide and Euthanasia Debated*, Michael M. Uhlmann (ed.), W.B. Eardmanns Publishing Co., Michigan, 1998

7. *Long Term Care: In an Ageing Society, Choices and Challenges for the '90s*, Gerald A. Larue and Rich Bayly, Prometheus Books, New York, 1992
8. *Time of Our lives*, Tom Kirkwood, Weidenfield and Nicholson, London, 1999

11
Quantum Theology: The Evolution of Faith

It is conclusions such as those of Professor Veatch presented in Chapter 10 on the critical issue of the definition of life and death that make me believe in quantum theology. As presented earlier in Chapters 1 and 2, quantum theology is a theology of acceptance — acceptance of uncertainty. Science and technology provide opportunity, but they do not eliminate uncertainty. Since we embrace the opportunity we must also accept the uncertainty. What does it mean to accept uncertainty? What is the doctrine of quantum theology?

It means we must recognise that new technological developments, such as IVF, intensive care units, implants, transplants and genetic engineering, all have their risks as well as their rewards. All have costs, personal and financial, and the costs are usually unpredictable.

It means we have to set priorities. We have to increase our awareness of alternatives and make conscious decisions. We need to speak out and require elected leaders to address the legal and financial issues of the length of life and the quality of life based upon a foundation rooted in faith and trust in God, rather than based upon political expediency.

Most importantly, acceptance of uncertainty also means acceptance of each other. It means acceptance of diversity.

None of us is right all the time, although sometimes it is difficult to admit it. In all these important issues, from the questions regarding the origins of the universe and life to the concerns over the cost and availability of health care to the debate over when life begins and ends, the experts agree there is uncertainty. Our religious heritage tells us we should accept one another as loving, caring persons. That means accepting a diversity of opinions. It means accepting that only God knows all the answers, that only God knows what is truly right or wrong in these areas.

It appears to be our destiny to search for the answers. Quantum theology tells us that it is also our destiny never to know for sure whether the answer is right or wrong. It also tells us sometimes we may not even have the right question.

We come to end of this odyssey, the end of this voyage. We also come back to the beginning. As Jesus said, "I am the alpha and the omega." The beginning and the end.

Thus we too come to the end and also back to the beginning. As we experience the mystery of birth we all must also expect the mystery of death; the transformation of the beauty of being into the uncertainty of dying.

As I walk on the seashore I contemplate these mysteries of experience, sand, sea and sky. They have always been here, immortal, and will continue to be here long past our coming and our going. Their destiny is to exist.

What is our destiny? What is our reason for being? Is it also only to be? Or is it to ask why? If it is, then our present emphasis on science and technology is perhaps misplaced. Science explores how and technology exploits the answer.

Neither science nor technology considers "*Why?*" to be relevant. Previously, even theology considered "Why?" to be irreverent.

Both positions are wrong. We must learn from both the past and the present, from philosophy and physics, that "Why?" is as important as "How?". We must accept that there may be no answer to either question. We must accept that the essence of nature is uncertainty. We must accept that the essence of being is unclear.

We sense that there is an essence beyond ourselves. But, we do not know, we do not understand and we often do not comprehend the origin of that presence. Our lack of understanding, our lack of description, our lack of even defining that presence, does not make it any less real. It just makes it unscientific.

However, we must never forget that measuring something, that explaining something with mathematical formulae, that performing experiments, is not an act of creation. It is instead often an act of self-delusion. The world and all that is in it was created before man and will continue to exist after man. Our description of the world and our efforts to explain it are part of the beauty of being. But, our efforts bear a pale resemblance to the completeness of being.

All too often, in the process of describing and defining nature and man's relation to nature, we impose our limits. Such limits create fragments. We see the world as "through a glass, darkly".

We also see the world as through a prism. As the prism separates light into colours, science often separates nature into fragments of reality. We are taught that assembling these fragments will create a whole. This teaching is wrong. It is wrong because the ways we perceive and measure nature are continually changing. Our efforts to synthesise them into an integral understanding are forever compromised by the intrinsic uncertainty of both the fragments and us, the synthesisers.

A consequence of this uncertainty inherent in science, and in knowledge itself, is an erosion of faith. Faith in science and technology all too easily supplements our faith in an Almighty God. However, faith in science is faith that the answers to "How?" will provide fulfilment.

Faith in science usually precludes even seeking "Why?" This is a tragic trade-off. The essence of being transcends the mechanical. It transcends the biological. The beauty of being encompasses mystery and uncertainty. We do not know how or why we are here. I believe we never will know in this life the answers to these metaphysical questions. We may know in a life hereafter. Or we may never know. That is both the mystery and the majesty of life.

Accepting this uncertainty, this quantum theology, does not mean accepting fate without question or effort.

However, it does mean that we must not demand answers. We must not demand that life treat each and every one alike and seek to ensure such uniformity. The essence of life, of being, as we know it, is diversity. We must learn to revel in and extol diversity rather than condemn it and attempt to obliterate it. Uncertainty appears to be the essence of our being. Change and diversity appear to be our destiny.

Epilogue

SIMPLY ASSISTING GOD

I am a humble artist
Moulding my earthly clod,
Adding my labour to nature's,
Simply assisting God.

Not that my effort is needed;
Yet somehow, I understand,
My maker has willed it that I too should have
Unmoulded clay in my hand.

— *Piet Hein, Grooks 1 (Doubleday, New York, N.Y., 1969)*

Index